Communications
in Computer and Information Science

Editorial Board Members

Joaquim Filipe ⓘ, *Polytechnic Institute of Setúbal, Setúbal, Portugal*
Ashish Ghosh ⓘ, *Indian Statistical Institute, Kolkata, India*
Lizhu Zhou, *Tsinghua University, Beijing, China*

Rationale
The CCIS series is devoted to the publication of proceedings of computer science conferences. Its aim is to efficiently disseminate original research results in informatics in printed and electronic form. While the focus is on publication of peer-reviewed full papers presenting mature work, inclusion of reviewed short papers reporting on work in progress is welcome, too. Besides globally relevant meetings with internationally representative program committees guaranteeing a strict peer-reviewing and paper selection process, conferences run by societies or of high regional or national relevance are also considered for publication.

Topics
The topical scope of CCIS spans the entire spectrum of informatics ranging from foundational topics in the theory of computing to information and communications science and technology and a broad variety of interdisciplinary application fields.

Information for Volume Editors and Authors
Publication in CCIS is free of charge. No royalties are paid, however, we offer registered conference participants temporary free access to the online version of the conference proceedings on SpringerLink http://link.springer.com by means of an http referrer from the conference website and/or a number of complimentary printed copies, as specified in the official acceptance email of the event.

CCIS proceedings can be published in time for distribution at conferences or as post-proceedings, and delivered in the form of printed books and/or electronically as USBs and/or e-content licenses for accessing proceedings at SpringerLink. Furthermore, CCIS proceedings are included in the CCIS electronic book series hosted in the SpringerLink digital library at http://link.springer.com/bookseries/7899. Conferences publishing in CCIS are allowed to use Online Conference Service (OCS) for managing the whole proceedings lifecycle (from submission and reviewing to preparing for publication) free of charge.

Publication process
The language of publication is exclusively English. Authors publishing in CCIS have to sign the Springer CCIS copyright transfer form, however, they are free to use their material published in CCIS for substantially changed, more elaborate subsequent publications elsewhere. For the preparation of the camera-ready papers/files, authors have to strictly adhere to the Springer CCIS Authors' Instructions and are strongly encouraged to use the CCIS LaTeX style files or templates.

Abstracting/Indexing
CCIS is abstracted/indexed in DBLP, Google Scholar, EI-Compendex, Mathematical Reviews, SCImago, Scopus. CCIS volumes are also submitted for the inclusion in ISI Proceedings.

How to start
To start the evaluation of your proposal for inclusion in the CCIS series, please send an e-mail to ccis@springer.com.

Marcelo Naiouf · Laura De Giusti ·
Franco Chichizola · Leandro Libutti
Editors

Cloud Computing, Big Data and Emerging Topics

12th Conference, JCC-BD&ET 2024
La Plata, Argentina, June 25–27, 2024
Revised Selected Papers

Editors
Marcelo Naiouf
III-LIDI
National University of La Plata
La Plata, Argentina

Laura De Giusti
III-LIDI
National University of La Plata
La Plata, Argentina

Franco Chichizola
III-LIDI
National University of La Plata
La Plata, Argentina

Leandro Libutti
III-LIDI
National University of La Plata
La Plata, Argentina

ISSN 1865-0929   ISSN 1865-0937 (electronic)
Communications in Computer and Information Science
ISBN 978-3-031-70806-0   ISBN 978-3-031-70807-7 (eBook)
https://doi.org/10.1007/978-3-031-70807-7

© The Editor(s) (if applicable) and The Author(s), under exclusive license to Springer Nature Switzerland AG 2025

This work is subject to copyright. All rights are solely and exclusively licensed by the Publisher, whether the whole or part of the material is concerned, specifically the rights of translation, reprinting, reuse of illustrations, recitation, broadcasting, reproduction on microfilms or in any other physical way, and transmission or information storage and retrieval, electronic adaptation, computer software, or by similar or dissimilar methodology now known or hereafter developed.
The use of general descriptive names, registered names, trademarks, service marks, etc. in this publication does not imply, even in the absence of a specific statement, that such names are exempt from the relevant protective laws and regulations and therefore free for general use.
The publisher, the authors and the editors are safe to assume that the advice and information in this book are believed to be true and accurate at the date of publication. Neither the publisher nor the authors or the editors give a warranty, expressed or implied, with respect to the material contained herein or for any errors or omissions that may have been made. The publisher remains neutral with regard to jurisdictional claims in published maps and institutional affiliations.

This Springer imprint is published by the registered company Springer Nature Switzerland AG
The registered company address is: Gewerbestrasse 11, 6330 Cham, Switzerland

If disposing of this product, please recycle the paper.

Preface

Welcome to the full paper proceedings of the 12th Conference on Cloud Computing, Big Data & Emerging Topics (JCC-BD&ET 2024), held in a hybrid setting (both on-site and live online participation modes were allowed). JCC-BD&ET 2024 was organized by the III-LIDI and the Postgraduate Office, both from the School of Computer Science of the National University of La Plata, Argentina.

Since 2013, this event has been an annual meeting where ideas, projects, scientific results and applications in cloud computing, big data and other related areas are exchanged and disseminated. The conference focuses on topics that allow interaction between academia, industry and other interested parties.

JCC-BD&ET 2024 covered the following topics: high-performance, edge and fog computing; internet of things; modeling and simulation; big and open data; machine and deep learning; smart cities; e-government; human-computer interaction; visualization; and special topics related to emerging technologies. In addition, special activities were also carried out, including 2 plenary lectures and 3 discussion panels.

In this edition, the conference received 37 submissions. The authors of these submissions came from the following 7 countries: Argentina, Cuba, Chile, the USA, Ecuador, India and Spain. All the accepted papers were peer-reviewed by at least three referees (single-blind review) and evaluated on the basis of technical quality, relevance, significance and clarity. To achieve this, JCC-BD&ET 2024 was supported by 52 Program Committee (PC) members and 31 additional external reviewers. According to the recommendations of the referees, 12 papers were selected for this book (32% acceptance rate). We hope readers will find these contributions useful and inspiring for their future research.

Special thanks to all the people who contributed to the conference's success: program and organizing committees, authors, reviewers, speakers and all conference attendees. Finally, we want to thank Springer for its support in publishing this book.

June 2024

Marcelo Naiouf
Franco Chichizola
Laura De Giusti
Leandro Libutti

Organization

General Chair

Marcelo Naiouf — Universidad Nacional de La Plata, Argentina

Program Committee Chairs

Armando De Giusti — Universidad Nacional de La Plata-CONICET, Argentina
Franco Chichizola — Universidad Nacional de La Plata, Argentina
Laura De Giusti — Universidad Nacional de La Plata and CIC, Argentina
Leandro Libutti — Universidad Nacional de La Plata and CIC, Argentina

Program Committee

María José Abásolo — Universidad Nacional de La Plata and CIC, Argentina
José Aguilar — Universidad de Los Andes, Venezuela
Jorge Ardenghi — Universidad Nacional del Sur, Argentina
Javier Balladini — Universidad Nacional del Comahue, Argentina
Oscar Bria — Universidad Nacional de La Plata and INVAP, Argentina
Silvia Castro — Universidad Nacional del Sur, Argentina
Mónica Denham — Universidad Nacional de Río Negro and CONICET, Argentina
Javier Diaz — Universidad Nacional de La Plata, Argentina
Ramón Doallo — Universidade da Coruña, Spain
Marcelo Errecalde — Universidad Nacional de San Luis, Argentina
Elsa Estevez — Universidad Nacional del Sur and CONICET, Argentina
Pablo Ezzatti — Universidad de la República, Uruguay
Aurelio Fernandez Bariviera — Universitat Rovira i Virgili, Spain
Fernando Emmanuel Frati — Universidad Nacional de Chilecito, Argentina
Carlos Garcia Garino — Universidad Nacional de Cuyo, Argentina

Carlos García Sánchez	Universidad Complutense de Madrid, Spain
Adriana Angélica Gaudiani	Universidad Nacional de General Sarmiento, Argentina
Graciela Verónica Gil Costa	Universidad Nacional de San Luis and CONICET, Argentina
Roberto Guerrero	Universidad Nacional de San Luis, Argentina
Waldo Hasperué	Universidad Nacional de La Plata and CIC, Argentina
Francisco Daniel Igual Peña	Universidad Complutense de Madrid, Spain
Tomasz Janowski	Gdańsk University of Technology, Poland
Laura Lanzarini	Universidad Nacional de La Plata, Argentina
Guillermo Leguizamón	Universidad Nacional de San Luis, Argentina
Edimara Luciano	Pontificia Universidade Católica do Rio Grande do Sul, Brazil
Emilio Luque Fadón	Universidad Autónoma de Barcelona, Spain
Mauricio Marín	Universidad de Santiago de Chile, Chile
Luis Marrone	Universidad Nacional de La Plata, Argentina
Katzalin Olcoz Herrero	Universidad Complutense de Madrid, Spain
José Angel Olivas Varela	Universidad de Castilla-La Mancha, Spain
Xoan Pardo	Universidade da Coruña, Spain
Patricia Pesado	Universidad Nacional de La Plata, Argentina
Mario Piattini	Universidad de Castilla-La Mancha, Spain
María Fabiana Piccoli	Universidad Nacional de San Luis, Argentina
Luis Piñuel	Universidad Complutense de Madrid, Spain
Adrian Pousa	Universidad Nacional de La Plata, Argentina
Marcela Printista	Universidad Nacional de San Luis, Argentina
Dolores Isabel Rexachs del Rosario	Universidad Autónoma de Barcelona, Spain
Nelson Rodríguez	Universidad Nacional de San Juan, Argentina
Juan Carlos Saez Alcaide	Universidad Complutense de Madrid, Spain
Aurora Sánchez	Universidad Católica del Norte, Chile
Victoria Sanz	Universidad Nacional de La Plata, Argentina
Remo Suppi	Universidad Autónoma de Barcelona, Spain
Francisco Tirado Fernández	Universidad Complutense de Madrid, Spain
Juan Touriño Dominguez	Universidade da Coruña, Spain
Gabriela Viale Pereira	Danube University Krems, Austria
Gonzalo Zarza	Globant, Argentina

Additional Reviewers

Hugo Alfonso
Javier Diaz
Enzo Rucci
Ivana Harari
César Estrebou
Alejandro Fernández
María Luján Ganuza
Mario Alejandra Garrido
Sergio Julián Grigera
Jorge Ierache
Martín Larrea
Antonio Lorenzo
Diego Martinez
Diego Montezanti
Joaquín Pina
Claudia Pons

Facundo Quiroga
Hugo Ramón
Andrés Rodriguez
Franco Ronchetti
Carolina Salto
Matías Selzer
Pablo Thomas
Diego Torres
Augusto Villa Monte
Paula Venosa
Virginia Ainchil
Pablo Tissera
Diego Encinas
Leandro Antonelli
Patricia Bazan

Contents

Parallel and Distributed Computing

Fast Genomic Data Compression on Multicore Machines 3
   Victoria Sanz, Adrián Pousa, Marcelo Naiouf, and Armando De Giusti

Machine and Deep Learning

Deep Learning-Based Instance Segmentation of Neural Progenitor Cell
Nuclei in Fluorescence Microscopy Images 17
   *Gabriel Pérez, Claudia Cecilia Russo, Maria Laura Palumbo,
and Alejandro David Moroni*

Object Recognition Models for Indoor Users' Location 30
   Franco M. Borrelli and Cecilia Challiol

CB-RISE: Improving the RISE Interpretability Method Through
Convergence Detection and Blurred Perturbations 45
   *Oscar Stanchi, Franco Ronchetti, Pedro Dal Bianco, Gastón Rios,
Santiago Ponte Ahon, Waldo Hasperué, and Facundo Quiroga*

Wavelength Calibration of Historical Spectrographic Plates with Dynamic
Time Warping 59
   *Santiago Andres Ponte Ahón, Juan Martín Seery, Facundo Quiroga,
Franco Ronchetti, Oscar Stanchi, Pedro Dal Bianco, Waldo Hasperué,
Yael Aidelman, and Roberto Gamen*

An Empirical Method for Processing I/O Traces to Analyze
the Performance of DL Applications 74
   *Edixon Parraga, Betzabeth Leon, Sandra Mendez, Dolores Rexachs,
Remo Suppi, and Emilio Luque*

Smart Cities and E-Government

Industry 5.0. Digital Twins in the Process Industry. A Bibliometric Analysis ... 93
   Federico Walas Mateo and Armando De Giusti

Visualization

An ABMS COVID-19 Propagation Model for Hospital Emergency
Departments 103
   *Morteza Ansari Dogaheh, Manel Taboada, Francisco Epelde,
Emilio Luque, Dolores Rexachs, and Alvaro Wong*

Emerging Topics

QuantumUnit: A Proposal for Classic Multi-qubit Assertion Development 121
   *Ignacio García-Rodríguez de Guzmán,
Antonio García de la Barrera Amo, Manuel Ángel Serrano,
Macario Polo, and Mario Piattini*

Tool for Quantum-Classical Software Lifecycle 132
   Jesús Párraga Aranda, Ricardo Pérez del Castillo, and Mario Piattini

Innovation in Computer Science Education

Strategies to Predict Students' Exam Attendance 145
   Gonzalo L. Villarreal and Verónica Artola

Computer Security

Prediction of TCP Firewall Action Using Different Machine Learning
Models 161
   *Amit Kumar Bairwa, Akshit Kamboj, Sandeep Joshi,
Pljonkin Anton Pavlovich, and Saroj Hiranwal*

Author Index 175

Parallel and Distributed Computing

Fast Genomic Data Compression on Multicore Machines

Victoria Sanz[1,2(✉)], Adrián Pousa[1], Marcelo Naiouf[1], and Armando De Giusti[1,3]

[1] III-LIDI, School of Computer Sciences, National University of La Plata, La Plata, Argentina
{vsanz,apousa,mnaiouf,degiusti}@lidi.info.unlp.edu.ar
[2] CIC, Buenos Aires, Argentina
[3] CONICET, Buenos Aires, Argentina

Abstract. Nowadays, Genomics has gained relevance since it allows preventing, diagnosing and treating diseases in a personalized way. The reduction in sequencing time and cost has increased the demand and, thus, the amount of genomic data that must be stored or transferred. Consequently, it becomes necessary to develop genome compression algorithms that help to reduce storage usage without consuming too much time. This is now possible thanks to modern multicore machines. This paper improves MtHRCM, a multi-threaded compression algorithm for large collections of genomes, by reducing its sequential component in order to enhance performance and scalability. Experimental results show that our optimized version is faster than MtHRCM and achieves the same compression ratio. Also, they reveal that this new version scales well when increasing the number of threads/cores for smaller test collections, while the high amount of simultaneous I/O requests to disk limits the scalability for larger test collections.

Keywords: Genomic Data Compression · Multi-threaded Hybrid Referential Compression Method · Multicore · Performance · DNA

1 Introduction

DNA sequencing (or genomic sequencing) allows determining the exact order of the nucleotide bases in the DNA of an organism. Analyzing the genomic information of an individual is of great interest since it determines not only physical characteristics but also the susceptibility to certain diseases, pharmacological compatibility, among others; consequently, the resulting information is used to prevent, diagnose and treat diseases in a personalized way [1–3].

Thanks to scientific and technological advances, the time and cost of sequencing have decreased rapidly, to the point that today it is possible to sequence an entire genome for less than USD 1000 in a period of 2 to 5 months [4,5]. However, the large amount of genomic data that must be stored for later use brings new challenges: reducing storage requirements without losing data and accelerating data transmission [6,7].

To tackle these challenges, several lossless compression algorithms for genomic sequences have emerged, which exploit the structure of these data in order to achieve a higher compression ratio than general-purpose compression algorithms (included in tools such as gzip, bzip2 and 7zip). In particular, these algorithms are classified into two categories, depending on whether they use references during compression or not: (i) reference-free algorithms compress the target sequence using only its internal characteristics; (ii) reference-based algorithms use one or more reference sequences to compress the target sequence, leveraging the high similarity between sequences of the same species [8–10].

Additionally, it is common to find collections of sequences that are stored together because they are related. Compared to compressing each sequence of the collection individually, batch compression is more efficient since it allows to carry out certain steps of the process only once (thus reducing compression time) and to obtain a higher compression ratio [11,12]. The FASTA format is widely used for storing genomic sequences, due to its simplicity and easy interpretation, and it is accepted by the most popular genomic databases and analysis tools [13–15].

Related to the previously mentioned, HRCM (Hybrid Referential Compression Method) [12] is a compression algorithm for collections of genomes in FASTA format, reference-based and lossless. For each to-be-compressed sequence, the algorithm performs a matching to find all the segments that are included in the reference sequence. As a result, a compressed sequence is obtained, which contains information about matches and mismatches. Additionally, if the user wishes so, first-matching results go through a second matching, where some already compressed sequences (prior to the current one) are taken as references. Finally, the file with the results is compressed with 7-zip.

The same authors proposed MtHRCM [16], a multi-threaded implementation of HRCM. The algorithm is built on the fact that the compression of most sequences is independent and can be carried out in parallel. Although MtHRCM improves the performance of HRCM, it achieves a poor speedup and does not scale well. This is mainly due to three factors: (i) the sequences that will be reference during the second matching are solved sequentially first; (ii) to guarantee correctness, each already compressed sequence is saved into a separate file and once the whole process is completed, these files are written into a new file sequentially in order; (iii) the contention at the I/O system.

In summary, it is of great importance to optimize the compression of genomic sequences in FASTA format, i.e. reducing compression time and maximizing compression ratio. In this paper, we propose MtHRCM-opt, an optimized version of MtHRCM that reduces its sequential component in order to enhance performance and scalability. Our experimental results show that MtHRCM-opt improves the performance of MtHRCM while maintaining the compression ratio. Also, the results reveal that MtHRCM-opt scales well when increasing the number of threads/cores for smaller test collections, while the high amount of simultaneous I/O requests to disk limits the scalability for larger test collections.

The rest of the paper is organized as follows. Section 2 summarizes the HRCM and MtHRCM algorithms. Section 3 describes our proposal. Section 4 shows our experimental results. Finally, Sect. 5 presents the main conclusions and future research.

2 Background

This section describes the HRCM and MtHRCM compression algorithms. Both algorithms compress a collection of FASTA sequences with a lossless reference-based approach.

2.1 HRCM Algorithm

HRCM (Hybrid Referential Compression Method) [12] consists of three processes: *extraction*, *matching* and *encoding*.

First, the algorithm *extracts* the reference sequence, that is, it keeps all the nucleotides (A, C, G, T) after converting them to uppercase and saves information about the segments originally in lowercase (position and length). Then, it constructs a hash table based on the extracted reference sequence, which stores for each possible k-mer (substring of k nucleotides) all its locations in the reference sequence or -1 if it does not exist. This table is used in the matching step.

Next, the *matching* process applies the following three steps to each to-be-compressed sequence:

- *Extraction:* similar to the extraction step of the reference sequence, but also it records information about: the stream identifier, the line width, 'N' characters and special characters.
- *First-level matching:* this step iterates over the extracted sequence using a sliding window of length k. When the k-mer currently in the window appears in the reference sequence, the position and length of the longest match are recorded in the results. Otherwise, the first base of the k-mer is recorded as a mismatched character. Then, the window is slided. This iteration continues until reaching the end of the sequence. Finally, if the sequence will be a reference during the second matching, the results of this step along with a hash table built from their entities (taken in pairs) are stored in memory. These data structures are used in the next step. The number of sequences that will be reference during the second matching (*refSeqNum*) is configurable.
- *Second-level matching:* similar to previous step, the results of the first-level matching are compressed, using a sliding window of length 2 and taking as references some of the *already compressed sequences* whose data structures are stored in memory (from the first one to the previous to this one, without exceeding *refSeqNum*). From the entities in the window, the longest match

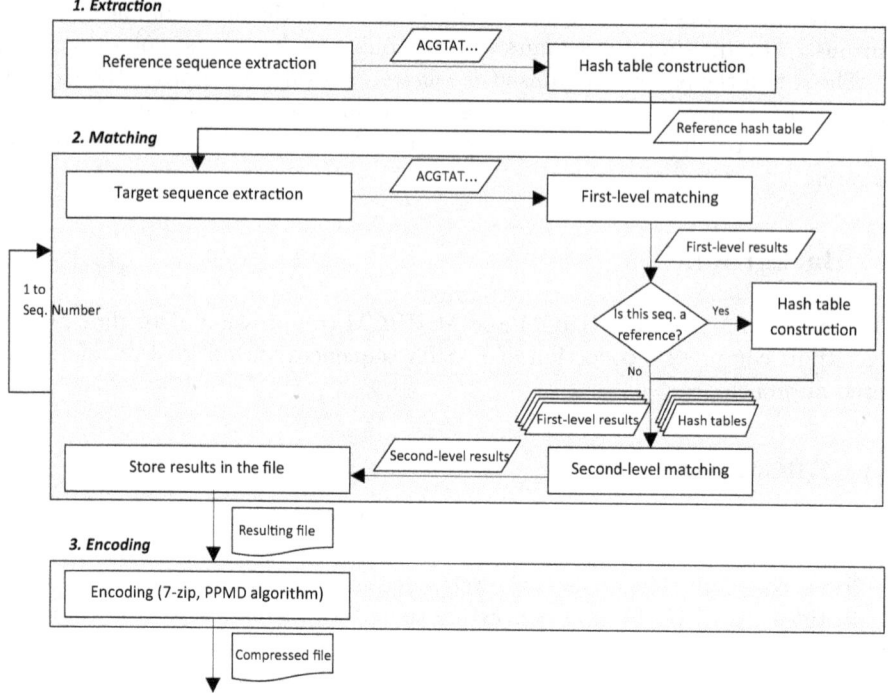

Fig. 1. HRCM algorithm

among all references is found and the information about the matched segment (sequence id, position, length) is written in the output. Otherwise, if a mismatch occurs, the first entity in the window is written in the output.

Finally, the output is *encoded* (compressed) with a general-purpose compressor (7-zip, PPMD algorithm). Figure 1 illustrates the whole algorithm.

The decompression procedure consists of decompressing the file with 7-zip and reconstituting each genomic sequence, respecting the order in which they were compressed and using the same value of *refSeqNum*. For that, both the compressor and the decompressor receive as input a file with the names of the sequences in the collection and the value of *refSeqNum*.

2.2 MtHRCM Algorithm

MtHRCM (Multi-thread HRCM) [16] uses multiple threads to compress the collection of sequences.

First, the main thread extracts the reference sequence and builds the associated hash table. Then, it processes those sequences that will be references during the second-level matching sequentially (one at a time). That is, the thread extracts the sequence and applies the first-level and second-level matching. Sequentiality ensures that when compressing a new sequence, all the data structures needed for the second-level matching are already stored in memory (i.e., the results of the first-level matching and their associated hash tables, related to those sequences to be taken as references). Each compressed sequence is written to a separate intermediate file.

Next, the remaining sequences are compressed in parallel. For that, the algorithm uses a pool of threads. Each thread takes one sequence at a time (dynamically) and processes it, saving the compressed output in a separate intermediate file.

After all sequences were compressed, the intermediate files are written sequentially (in order) into the final output file, which is then compressed with the general-purpose compressor (7-zip).

3 Our Proposal: MtHRCM-Opt Algorithm

Our proposal (MtHRCM-opt algorithm) focuses on improving the performance and scalability of MtHRCM by reducing the sequential parts of the algorithm. Specifically, at the start of execution, the sequences that will be references during the second-level matching are processed serially. Additionally, at the end of execution, the step of combining the intermediate files with compressed sequences into a single file is also serial. Consequently, MtHRCM loses parallelism when the time of these parts increases. This occurs when more sequences are taken as references for the second-level matching and when more to-be-compressed sequences are included in the collection, respectively.

MtHRCM-opt uses multiple threads to compress the collection of sequences.

First, the main thread extracts the reference sequence and builds the associated hash table. Then, a pool of threads will start to work on the to-be-compressed sequences.

Each thread dynamically picks the next sequence from the collection and processes it in the following way. The thread completes the first-level matching, which has no data dependence. Then, it waits for all the data structures required by the second-level matching to be ready (i.e., the first-level matching results and their associated hash tables, related to those sequences to be taken as references). Specifically, let i be the index of the sequence being processed, the thread must wait for the data structures of sequences with index between 1 and $minimum(i-1, refSeqNum)$ to be ready. Once this condition is met, the thread completes the second-level matching and stores the results in a separate intermediate file.

After all sequences were processed, the intermediate files are compressed with 7-zip.

In this way, our implementation parallelizes the computation of the sequences that are references during the second-level matching and eliminates the single file creation step at the end of execution. The latter change involves modifying slightly the decompressor because its input is now divided into several files (i.e., one per to-be-decompressed sequence).

4 Experimental Results

Our experimental platform is a machine composed of two Intel Xeon E5-2695 v4 processors, 128 GB RAM and a SAS disk. Each processor has eighteen 2.10 Ghz cores, thus the machine has thirty-six cores in total. Hyper-Threading and Turbo Boost were disabled. All the to-be-compressed sequences are stored on the local disk.

Tests were performed with human genomes. Specifically, we considered a collection of 1100 to-be-compressed genomes, out of which 1092 are from the 1000 Genome Project, 5 are from the UCSC Genome Browser (HG16, HG17, HG18, HG19 and HG38), 2 are the Korean genomes KOREF_20090131 and KOREF_20090224, and the last is the HuRef genome [17–21]. We used the UCSC HG13 genome as reference [18]. Human genomes contain 24 chromosomes (identified as 1, 2, .., 22, X, Y) and have a size of ∼3000 MB each.

Each test considered a specific chromosome, for example: the tests on chromosome 1 use the chromosome 1 of HG13 as reference and a to-be-compressed collection of 1100 chromosomes 1 (taken from the above-mentioned genomes). This grouping allows the compressor to leverage the similarity between the to-be-compressed sequences and the reference. Moreover, it allows us to evaluate the scalability of the algorithm, as explained next.

Table 1 shows for each test collection: the total size, the average sequence size, and the total compressed size by HRCM (all in MB). There are 24 test collections in total, each one corresponds to a particular chromosome. Observe that in general, the smaller the chromosome ID, the larger the collection size. Also, each test collection includes sequences of similar size (near to the average). Thus, larger test collections are composed of larger sequences, and smaller test collections have smaller sequences.

To prove the effectiveness of our proposal, we compare HRCM (sequential code), MtHRCM (original parallel code) and MtHRCM-opt (our parallel code). In the latter two cases, the runs were carried out with 4 and 8 threads/cores. Later, in order to evaluate the scalability of MtHRCM-opt, more experiments were run with 16 and 32 threads/cores. The number of references for the second-level matching (*refSeqNum*) was set to 275, which corresponds to 25 percent of the to-be-compressed collection.

We evaluate three metrics: (1) Compression ratio, which is the ratio between the uncompressed data size and the compressed data size, to quantify the effectiveness of the compressor; (2) Speedup, defined as the ratio between the execution time of the sequential algorithm and the execution time of the parallel algorithm, to evaluate the improvement in performance when increasing the number of cores and the problem size; (3) Throughput (in MB/s), calculated as the ratio between the uncompressed data size (in MB) and the compression time (in seconds), to show the amount of data processed per unit time.

This article focuses on the compression process, since it is the most time-consuming operation.

First, we study the compression effectiveness of our proposal. In each test performed (chromosomes/threads), MtHRCM-opt obtained the same compressed

Table 1. Test collections used

# Chr	Total size (MB)	Avg. sequence size (MB)	Compressed size (MB) by HRCM
chr1	264921.16	240.84	104.39
chr2	258642.66	235.13	108.69
chr3	210601.02	191.46	96.09
chr4	203091.28	184.63	88.86
chr5	192408.64	174.92	80.04
chr6	181929.30	165.39	75.13
chr7	169205.92	153.82	71.14
chr8	155599.61	141.45	73.24
chr9	150089.29	136.44	53.37
chr10	144151.52	131.05	58.02
chr11	143542.08	130.49	60.76
chr12	142264.24	129.33	67.64
chr13	122466.04	111.33	40.11
chr14	114095.71	103.72	37.68
chr15	109019.07	99.11	36.72
chr16	96085.52	87.35	43.28
chr17	86312.02	78.47	42.54
chr18	83039.52	75.49	33.11
chr19	62833.48	57.12	28.27
chr20	67014.55	60.92	26.01
chr21	51146.38	46.50	17.31
chr22	54476.50	49.52	18.01
chrX	165133.60	150.12	54.97
chrY	30615.47	27.83	2.60
Sum	3258684.59	∼3000	1318.00

size as HRCM and MtHRCM. These values are shown in Table 1. This behavior is expected since the compression methodology is identical for the three algorithms. In general, all test collections (a total of 3258684 MB or ∼3 TB) were compressed to 1318 MB, giving a compression ratio of ∼2472. That is, compressed data require 2472 times less storage than the original data.

Next, we verify the performance improvement of our algorithm. Figure 2 compares the speedup of MtHRCM and MtHRCM-opt, for all chromosomes, with 4 and 8 threads/cores. As can be seen, MtHRCM achieves a poor speedup and does not scale well when increasing the number of cores from 4 to 8. In contrast, MtHRCM-opt exhibits good speedup values and outperforms MtHRCM in all cases.

Then, we investigate the scalability of MtHRCM-opt. Figure 3 shows the speedup of MtHRCM-opt, for all chromosomes, when the number of threads/-

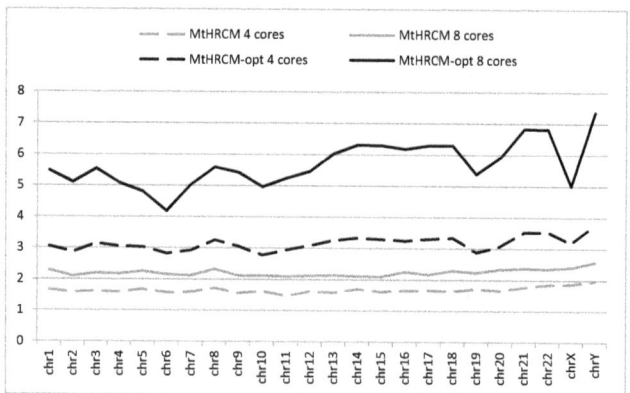

Fig. 2. Speedup comparison between MtHRCM and MtHRCM-opt

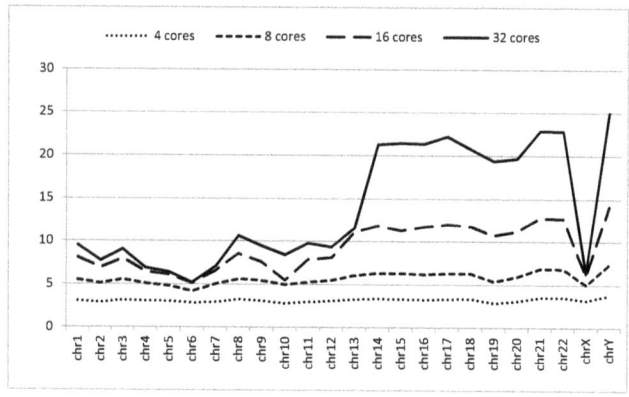

Fig. 3. Speedup of MtHRCM-opt when increasing the number of cores

cores is further increased (4, 8, 16 and 32). Here we can see two groups of chromosomes {1..13, X} and {14..22, Y}. Although MtHRCM-opt exhibits a good speedup with 4 and 8 cores (for all chromosomes), when the number of cores becomes larger the speedup increases very slightly for the first group of chromosomes, while for the second group the gain in speedup is more pronounced.

To determine the cause of this behavior, we examined the overhead introduced by parallelism in each execution. The overhead is a percentage calculated with the formula $Ov = \frac{T_{ov} \cdot 100}{T_p}$, where $T_{ov} = T_p - \frac{T_s}{cores}$, T_p is the execution time of the parallel algorithm and Ts is the execution time of the sequential algorithm. In an ideal scenario where there is no overhead due to parallelism $T_p = \frac{T_s}{cores}$, giving an overhead of 0.

Figure 4 illustrates the overhead for each test of MtHRCM-opt (chromosomes, threads). The figure reveals that, for a fixed chromosome, the overhead grows as the number of cores increases. Also, it shows that the chromosomes of the first

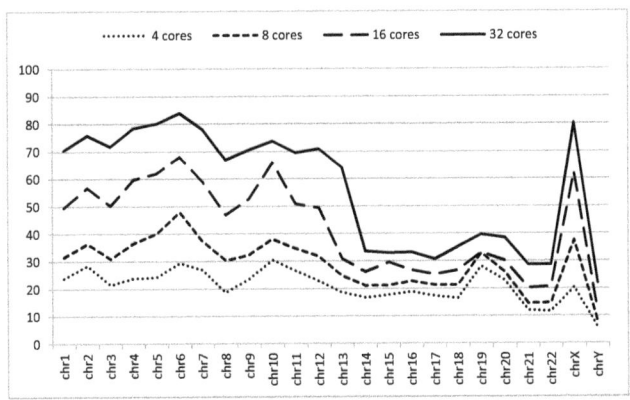

Fig. 4. Overhead of MtHRCM-opt

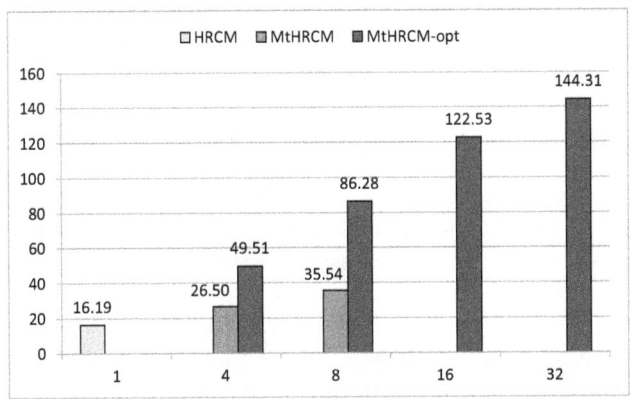

Fig. 5. Overall Throughput (in MB/s) of HRCM, MtHRCM and MtHRCM-opt

group have a greater overhead than those of the second group. The main source of overhead in our algorithm is disk contention. Disk I/O is performed to extract the information of each sequence from file and write the compressed sequence to disk. For a fixed test collection, as more threads are involved in compression, more sequences are processed in parallel, therefore there will be more I/O requests that the disk must serve simultaneously. This results in disk contention, which causes long latencies. This overhead is even greater for collections with larger sequences (like those of the first group) since they require more I/O operations.

From these results we conclude that MtHRCM-opt scales well when increasing the number of threads/cores for smaller test collections, while the high amount of simultaneous I/O requests to disk limits the scalability for larger test collections.

Finally, we analyze the overall throughput of the three algorithms to provide a more concrete interpretation of the presented results. Figure 5 shows the overall throughput (in MB/s) of HRCM, MtHRCM and MtHRCM-opt. These values were calculated considering the total size and the total compression time of all test collections (24 chromosomes), for each algorithm/cores. As can be derived from these results, on our platform HRCM compresses all data (∼3258684 MB) in ∼56 h, MtHRCM takes 34.16 and 25.47 h (with 4 and 8 cores respectively), while MtHRCM-opt completes the compression in 18.28, 10.49, 7.39 and 6.27 h (with 4, 8, 16 and 32 cores respectively). In other terms, per human genome (∼3000 MB) HRCM uses about 185 s, MtHRCM uses 113 s and 84 s (with 4 and 8 cores respectively), while MtHRCM-opt uses 60 s, 35 s, 24 s, and 21 s (with 4, 8, 16 and 32 cores respectively).

5 Conclusions and Future Work

In this paper we presented MtHRCM-opt, an optimized version of the Multi-thread Hybrid Referential Compression Method (MtHRCM), which allows compressing large collections of genomes on multicore machines. The optimizations were applied in order to reduce the sequential components of the original parallel code, which result in poor performance and scalability.

Our experimental results showed that MtHRCM-opt improves the performance of MtHRCM while maintaining the compression ratio. Also, they reveal that MtHRCM-opt scales well when increasing the number of threads/cores for smaller test collections, while the high amount of simultaneous I/O requests to disk limits the scalability for larger test collections.

As a main concrete result, on our platform with 4 and 8 cores, MtHRCM-opt compressed the whole collection of 1100 human genomes (∼3258684 MB) in 18.28 and 10.49 h respectively, while MtHRCM did the same in 34.16 and 25.47 h respectively. Furthermore, when using 16 and 32 cores, MtHRCM-opt completed the compression in 7.39 and 6.27 h.

In future, we plan to adapt HRCM to work on multicore clusters. Also, we aim to compare the performance obtained with that of MtHRCM-opt and HadoopHRCM (a version of HRCM suitable for compressing large collections of genomes stored in a cloud environment).

References

1. NHS England: Health Education England's Genomics Education Programme: what is genomics?. https://www.genomicseducation.hee.nhs.uk/education/core-concepts/what-is-genomics/
2. National Research Council: Mapping and Sequencing the Human Genome, p. 1988. The National Academies Press, Washington, DC (1988)
3. Drew, L.: Pharmacogenetics: the right drug for you. Nature **537**, S60–S62 (2016)
4. Wetterstrand KA.: DNA Sequencing Costs: Data from the NHGRI Genome Sequencing Program (GSP). www.genome.gov/sequencingcostsdata
5. National Human Genome Research Institute: Frequently Asked Questions and Resources. https://www.genome.gov/Clinical-Research/Secondary-Genomics-Findings-Service/FAQ-Resources

6. National Human Genome Research Institute: Genomic Data Science. https://www.genome.gov/about-genomics/fact-sheets/Genomic-Data-Science
7. Stephens, Z.D., et al.: Big Data: Astronomical or Genomical? PLoS Biol. **13**(7), e1002195 (2015)
8. Kredens, K.V., et al.: Vertical lossless genomic data compression tools for assembled genomes: a systematic literature review. PLOS One **15**(5), e0232942 (2020)
9. Hosseini, M., et al.: A survey on data compression methods for biological sequences. Information **7**(4), 56 (2016)
10. Wandelt, S., et al.: Trends in genome compression. Curr Bioinf. **9**(3), 315–326 (2013)
11. Deorowicz, S., et al.: GDC 2: compression of large collections of genomes. Sci. Rep. **5**, 11565 (2015)
12. Yao H,. et al.: HRCM: an efficient hybrid referential compression method for genomic big data. BioMed. Res. Int. **2019**, Article ID 3108950 (2019)
13. Whitehoyse D., Rapley R.: Chapter 5: Introductory bioinformatics. In: Genomics and Clinical Diagnostics. Royal Society of Chemistry (2019)
14. Gebank: NIH genetic sequence database. https://www.ncbi.nlm.nih.gov/genbank/
15. Wheeler, D., Bhagwat, M.: BLAST QuickStart: example-driven web-based BLAST tutorial. Methods Mol. Biol. **395**, 149–176 (2007)
16. Yao, H., et al.: Parallel compression for large collections of genomes. Concurr. Comput. Pract. Exper. **34**(2), e6339 (2021)
17. The International Genome Sample Resource (IGSR). https://www.internationalgenome.org/
18. UCSC Genome Browser Group: University of California, Santa Cruz. http://genome.ucsc.edu
19. Ahn, S.M., et al.: The first Korean genome sequence and analysis: full genome sequencing for a socio-ethnic group. Genome Res. **19**(9), 1622–1629 (2009)
20. KOBIC: Korea Bioinformation Center. ftp://ftp.kobic.kr/pub/KOBIC-KoreanGenome/
21. The National Center for Biotechnology Information, U.S.: Genome assembly HuRef. https://www.ncbi.nlm.nih.gov/datasets/genome/GCF_000002125.1/

Machine and Deep Learning

Deep Learning-Based Instance Segmentation of Neural Progenitor Cell Nuclei in Fluorescence Microscopy Images

Gabriel Pérez[1,2(✉)], Claudia Cecilia Russo[2], Maria Laura Palumbo[1,3], and Alejandro David Moroni[1,3]

[1] Centro de Investigaciones y Transferencia del Noroeste de la Provincia de Buenos Aires (CIT NOBA)-UNNOBA-UNsADA-CONICET, Junín, Buenos Aires, Argentina
`gabriel.perez@itt.unnoba.edu.ar, {mlpalumbo, admoroni}@comunidad.unnoba.edu.ar`
[2] Instituto de Investigación y Transferencia en Tecnología (ITT), Universidad Nacional del Noroeste de la provincia de Buenos Aires (UNNOBA), Junín, Buenos Aires, Argentina
`claudia.russo@itt.unnoba.edu.ar`
[3] Centro de Investigaciones Básicas y Aplicadas (CIBA), Junín, Buenos Aires, Argentina

Abstract. In this work, a Deep Learning-based machine vision model was developed for the detection, segmentation and counting of Neural Progenitor Cell nuclei from fluorescence microscopy images. The cells were obtained from adult mice and cultivated in vitro, with cellular nuclei labeled using DAPI dye. Convolutional neural networks for instance segmentation, specifically the Mask R-CNN model with ResNet-50 and ResNet-101 backbones, were trained to recognize the nuclei, and their results were evaluated. Nuclei labeling was implemented semi-automatically, applying a Superpixel technique and then refining the segmentations from a manual process, also using a pre-trained model, which allowed to assemble a dataset of 66 images with 6392 labels in total. The results obtained with the Resnet-50 backbone show that there is an effectiveness of 98.6% for between the specialist count and model-predicted count, in addition to having an mAP50 of 98.0%. This approach has the potential to significantly reduce the time and effort required to analyze large image sets, which is especially useful in studies that require repetitive and detailed cellular analysis.

Keywords: Instance segmentation · Deep Learning · Fluorescence Microscopy · Cell Nuclei

1 Introduction

In the last few years, there has been significant progress regarding the acquisition of large sets of microscopy images in short periods of time, enabling computational analysis [1]. This paves the way for a series of tools capable of obtaining automatic information and supporting decision-making in the field of bioimaging [2, 3]. Some procedures carried out in laboratories are long, exhaustive, and repetitive, where the automation of different processes has become essential. In this context, digital image analysis plays an important role in studying patterns relevant to specialists [4].

In the study of the hippocampus, a brain area related to learning and memory, there are Neural Progenitor Cells (NPC). These cells proliferate and then differentiate into three types of cells in the central nervous system: neurons, astrocytes, and oligodendrocytes. NPC from adult mice can be cultivated in vitro and studied through fluorescence microscopy images, which involves complex processes that could be automated using Deep Learning (DL) models.

Several approaches can be found in the literature about the automation through digital image analysis, for example, articles that provide a response to the characterization of cellular markers through algorithms based on color, texture, shape, and contrast [5, 6]. Likewise, there are works that do so through the use of Artificial Intelligence (AI) with supervised and semi-supervised learning. Various works use different Machine Learning (ML) techniques [7]. Particularly, Deep Learning (DL) models built with convolutional neural networks are used to obtain feature patterns of the elements to be recognized and then perform a classification process, involving general applications [8], cellular analysis [9, 10], and cellular tracking [11]. Among the techniques mentioned above, some focus on object segmentation [4, 11] and others focus on detection [11, 12]; both of which are an important first step for further analysis according to the specialist's needs.

ML methods are increasingly being implemented to characterize different markers in disease prognosis [13]. Through ML, disease biomarkers can be obtained without the need for a costly and invasive process, providing prognostic information similar to traditional techniques [2]. Specifically, there are pathology-dedicated softwares for the study of cell proliferation markers in renal cell carcinoma [3]. Image analysis alongside ML are tools of a great value for locating biomarkers; through a classifier, a heat map can be created to highlight tumor-infiltrating lymphocytes (TILs) and cancer cells data in an image, allowing for quantification controls in biomedical research and diagnosis [14].

In fluorescence microscopy imaging studies, fluorescent markers are commonly used to identify subcellular structures such as protein complexes, chromosomes, genes, and gene mutations [1]. To simplify the segmentation process, certain projects use fluorescent chromatin or DNA labels, where ML is widely used to classify cellular morphology obtained by fluorescent markers [15]. Particularly regarding the recognition and detection of cellular nuclei, there exist competitions like the Data Science Bowl 2018 [16], through which models like Mask-RCNN [17] could be trained to simultaneously detect and segment divergent nuclei from different sources to aid medical development and discovery.

An important step in the implementation of DL models is the assembly of the dataset and the corresponding labels. There are manual image labeling tools that allow to create bounding boxes, while others do so based on segments in the image, such as Superpixels [18]. It is indeed a laborious process that in certain areas of study can lead to errors if not done by the hand of a specialist in the field, potentially causing over-segmentation, underestimation, discard and add objects that are not of interest. These issues led to the research of new labeling methods based on models that automatically extract visual features, thus making the process easier. There exist articles where Superpixel models form an essential part of a classification process in medical images, such as in the segmentation of brain MRI images [19], or in optic nerve images [20] where segments are used to extract relevant features.

The present research aims to develop a model based on vision and artificial intelligence to automate processes in genetic laboratories. This model, intended to obtain data and relevant information about cellular markers and behaviors, will be applied to a dataset provided by expert geneticists. Specifically, this article shows the development of a model with the ability to detect, segment, and count cell nuclei in an image using a convolutional neural network called Mask-RCNN [17]. To achieve this, fluorescence microscopy images of NPCs cultivated in vitro from adult mice were used, where cellular nuclei were labeled with DAPI dye [21]. Additionally, this work presents a semi-automatic approach for generating nucleus labels, through the use of a Superpixel and a proprietary pre-trained model, followed by a manual process.

2 Image Acquisition and Processing

2.1 About the Scene

Images were acquired by geneticists in the laboratory through a process that involved:

1. Extraction of Neural Progenitor Cells (NPC) from the hippocampus of adult BALB/c mice.
2. Cell cultivation in the presence of growth factors.
3. Cell proliferation analysis using immunofluorescence, which stained cellular nuclei with a DAPI marker in blue.
4. Acquisition of images using a Nikon Eclipse E800 fluorescence microscope with a 40× objective (Fig. 1).

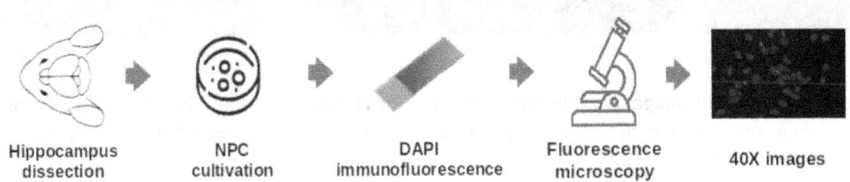

Fig. 1. DAPI image acquisition process. (Color figure online)

This process was performed by geneticists with the purpose of studying the hippocampus, an area of the brain related to learning and memory, where NPCs are found. These cells can proliferate and differentiate into three types of cells of the central nervous system: neurons, astrocytes, and oligodendrocytes. In order to study these behaviors, it was essential to analyze the presence of NPC nuclei through microscopic imaging.

In these images, we can see the presence of NPC nuclei, some of them separated and others overlapping. Additionally, there are small objects considered noise for the scene. The images have a resolution of 1280 × 960 pixels with the highest amount of data in the B (Blue) layer of the RGB color scheme.

2.2 Proposed Solution

The solution proposed in this work involves three stages, the first one consists in the preprocessing of DAPI cell nuclei images, where image enhancement techniques were used. The second stage consists of semi-automatic labeling of nuclei based on a Superpixel method and finally there is the training of an instance segmentation network of these nuclei.

Image Preprocessing and Enhancement. To begin with, the image was preprocessed in order to visually enhance it. The enhancement criteria were based on functions that can be applied to the image, such as contrast and brightness adjustment and noise attenuation. The purpose in mind was to highlight the objects of interest over the context in which they are located, thus creating contrast with other objects and the background.

In this step an enhancement was applied specifically in layer B (Blue) that composes the RGB of the color image, this was done because the image is purely blue and most of the data is in that layer. Contrast was achieved by applying a normalization technique only to this layer, considering its dynamic range, thus stretching the histogram (Fig. 2).

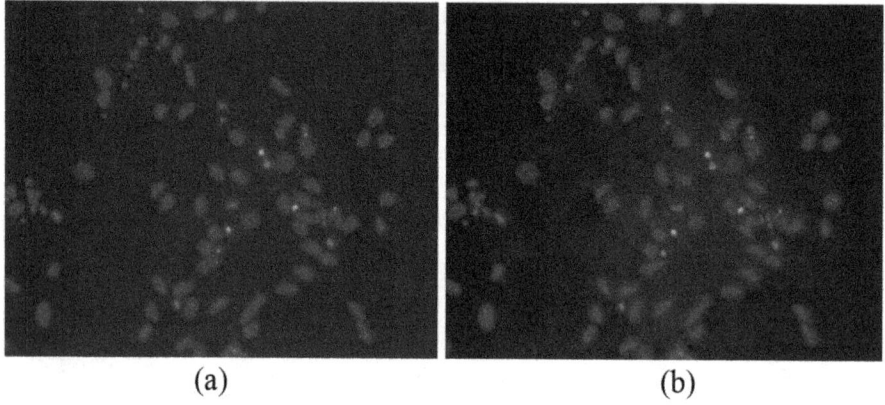

Fig. 2. (a) Original image. (b) Preprocessed image after normalization. The preprocessed image generates a contrast effect between the objects of interest and the background. (Color figure online)

Semi-automatic Labeling of Cellular Nuclei. On the preprocessed images, a set of labels was created to be used as input for the convolutional neural network. Given that the network to be used has the ability to detect objects in the image and also segment them, an approach to labeling that fits this capability was adopted. The inputs for this network are labels in segmentation or mask format, which is why this labeling methodology was approached.

Considering the characteristics of the image and the complexity of manually labeling cellular nuclei using polygons, a semi-automatic process based on a clustering algorithm was chosen. Specifically, the Felzenszwalb Superpixel algorithm [22] was used. This algorithm groups regions of the image that share similarities in terms of color and texture. As a result, several segments or superpixels were obtained that clearly separated the objects of interest from the surrounding background (see Fig. 3a).

To achieve a more effective isolation of the objects of interest from the background, it was considered that the background generally occupies the largest space in the image. Upon the set of segments or superpixels, the one with the largest area, corresponding to the background, was identified, and subsequently removed. This resulted in a clearer segmentation of the cell nuclei (see Fig. 3b).

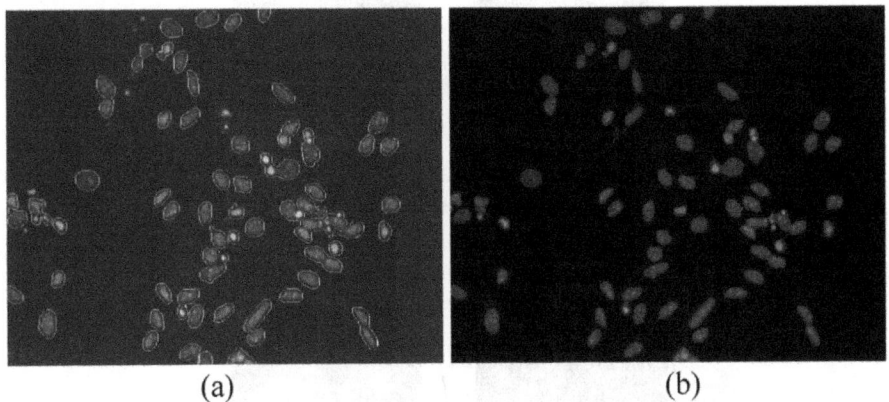

Fig. 3. (a) Preprocessed image of segments obtained from the Felzenszwalb Superpixel. (b) Image of segments obtained from Superpixel algorithm without the background. (Color figure online)

It is important to clarify that the algorithm relies on local contrast to identify the different elements in the image, that's why preprocessing was done before applying it. Figure 4 shows the result of applying it to the original image. It can be seen that there are areas where it couldn't detect some nuclei that are visible to the naked eye, this is due to the lack of contrast they present with the background.

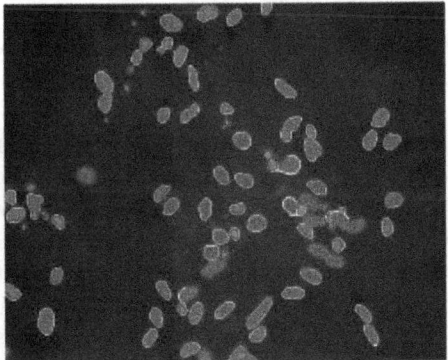

Fig. 4. Original image of segments obtained from the Felzenszwalb Superpixel.

Figure 3b shows that there are some superpixels that contain more than one cell nucleus. Object separation techniques were applied in these cases, but unfortunately, the results were not satisfactory due to the morphological variations that some of these nuclei may present. In some cases, the different nuclei were separated appropriately, but the segmentation obtained was wrong (Fig. 5).

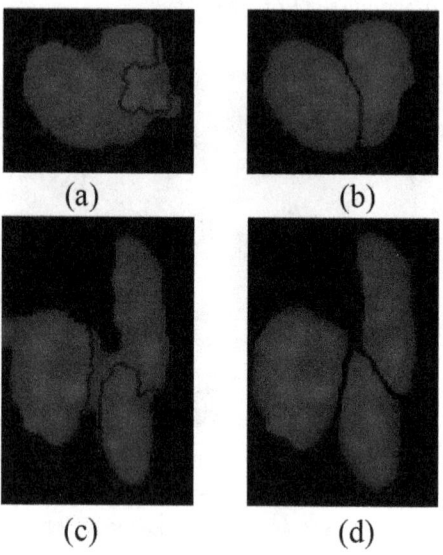

Fig. 5. Incorrect segmentation examples. (a) (c) Splitting with the Watershed algorithm. (b) (d) Desired splitting.

Considering that data is the raw material for deep learning models, incorrect segmentation does not faithfully represent the object. This can lead to errors in the training of the detection model and may distort the learning of a correct representation of the object. Based on this, it was chosen to manually separate connected superpixels of nuclei and remove segments that caused noise. This editing process was performed using a freehand drawing tool on the segmentation images, eliminating incorrect parts and separating nuclei as necessary. Each labeled image was supervised by a geneticist and adjusted if necessary, ensuring proper segmentation.

Figure 6.a shows the final image with manually modified superpixels. Figure 6.b shows the input labels for the convolutional neural network, each cellular nucleus bounded by a contour representing the segmentation and additionally a bounding box locating it in the image.

Applied techniques were able to generate labels to train with diverse data under different illumination and contrast conditions. The final dataset from these techniques allowed to generate 22 original and 22 preprocessed images, both sets with 1773 labeled nuclei, resulting in a total of 3546 labels for the entire dataset.

Convolutional Neural Network Training. The network used for object recognition in this work is a two-stage convolutional network based on a Feature Pyramid Network (FPN) architecture. Specifically, the Mask R-CNN architecture [23] was employed, which is an extension of the Faster R-CNN model. In addition to object detection, Mask

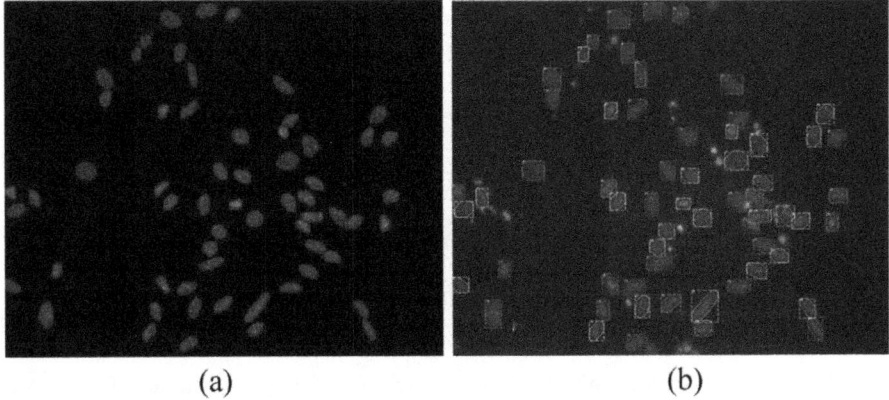

Fig. 6. (a) Image showing manually modified segments. (b) Image showing the input labels to the neural network.

R-CNN also provides segmentation of each detected object in the form of a binary mask. This technique, known as instance segmentation, allows to accurately distinguish the area of each object in the image, even in the presence of overlaps and occlusions.

Tests were performed on Mask R-CNN with ResNet-50 and ResNet-101 backbones [24]. ResNet-50 is a deep convolutional network with 50 layers that uses residual blocks to make training easier, while ResNet-101 is a deeper variant with 101 layers. The aim was to evaluate the impact of each of these backbones on this dataset, considering the difference in the network's ability to capture complex object features.

From preliminary tests using the semi-automatically obtained dataset and a Mask R-CNN neural network with ResNet-50, a model was obtained with the ability to recognize cell nuclei with an accuracy of 97.7%. Performance tests of the trained model were conducted with another dataset of 22 images, obtaining new labels to perform further training. In this case, the model was used as an automatic object labeller. Despite the high accuracy percentage, there were instances that went unrecognized or were misclassified. Therefore, small modifications were made to the model results, adding missing objects or removing incorrect ones, resulting in a new dataset.

The neural network training was then performed with the merging of the two datasets, resulting in a total of 66 images with 6392 labels. The images were split in groups of 75% for training and 25% for testing. All images used have both original and preprocessed versions as a data augmentation method to generate variability in the data environment. It is important to clarify that an original image and its preprocessed version are included in the same set, whether it's for training or testing, to minimize bias in both the training process and model testing, ensuring that there is no corruption between training and testing samples.

The original images have a resolution of 1280 × 980 pixels, which were resized to 1280 × 1280 using the square method before entering into the neural network. Training for both ResNet-50 and ResNet-101 was performed for a maximum of 50 epochs, with 100 iterations each, using pre-trained initial weights from the MS COCO dataset [25].

The best weights were selected based on the criterion of the lowest average percentage error between the actual count and the count predicted by the model on the test set.

The error was calculated as follows:

$$E = \frac{1}{n} \sum_{i=1}^{n} \left| \frac{N^i_{detected} - N^i_{real}}{N^i_{real}} \right| \times 100\% \quad (1)$$

where:

n: the number of images in the dataset.
$N^i_{detected}$: the number of objects detected by the model in image i.
N^i_{real}: number of real objects in image i.

Additionally, metrics such as mAP IoU = 0.5 (mAP50), precision, recall, F1-score, and other errors between real and predicted quantities by the model were considered.

3 Results and Discussion

The results obtained in the evaluation of the models for recognizing cellular nuclei of NPCs can be observed in Fig. 7, where the results for both the ResNet-50 and ResNet-101 backbones are shown. The graphs display a scatter plot that relates the quantities obtained by the model and the visual count, along with the regression line showing the correspondence between both variables.

The exact results are shown in Table 1, where different model evaluation metrics are shown. The effectiveness of the model is calculated as 100% minus the average percentage error:

$$Ef = 100\% - \frac{1}{n} \sum_{i=1}^{n} \left| \frac{N^i_{detected} - N^i_{real}}{N^i_{real}} \right| \times 100\% \quad (2)$$

where:

n: the number of images in the dataset.
$N^i_{detected}$: the number of objects detected by the model in image i.
N^i_{real}: number of real objects in image i.

Other metrics related to the actual count of cell nuclei to those predicted by the model were also analyzed, as shown in Table 2. These metrics include Mean Squared Error (MSE), Root Mean Squared Error (RMSE), Mean Absolute Error (MAE), Coefficient of Determination (R2), and also the maximum error between both counts.

The results of both architectures with the ResNet-50 and ResNet-101 backbones, show that in this context and for these images, a model with high depth of learning is not necessary. Cell nuclei exhibit similar characteristics in terms of texture, color, size, and shape, the challenge lies in the occlusion or overlap that they may present in different images. A complex model with a backbone like ResNet-101 may induce overfitting in this context.

It can be seen that both ResNet-101 and ResNet-50 have similar results in terms of mAP50 metrics and Effectiveness. This provides us with information about the results

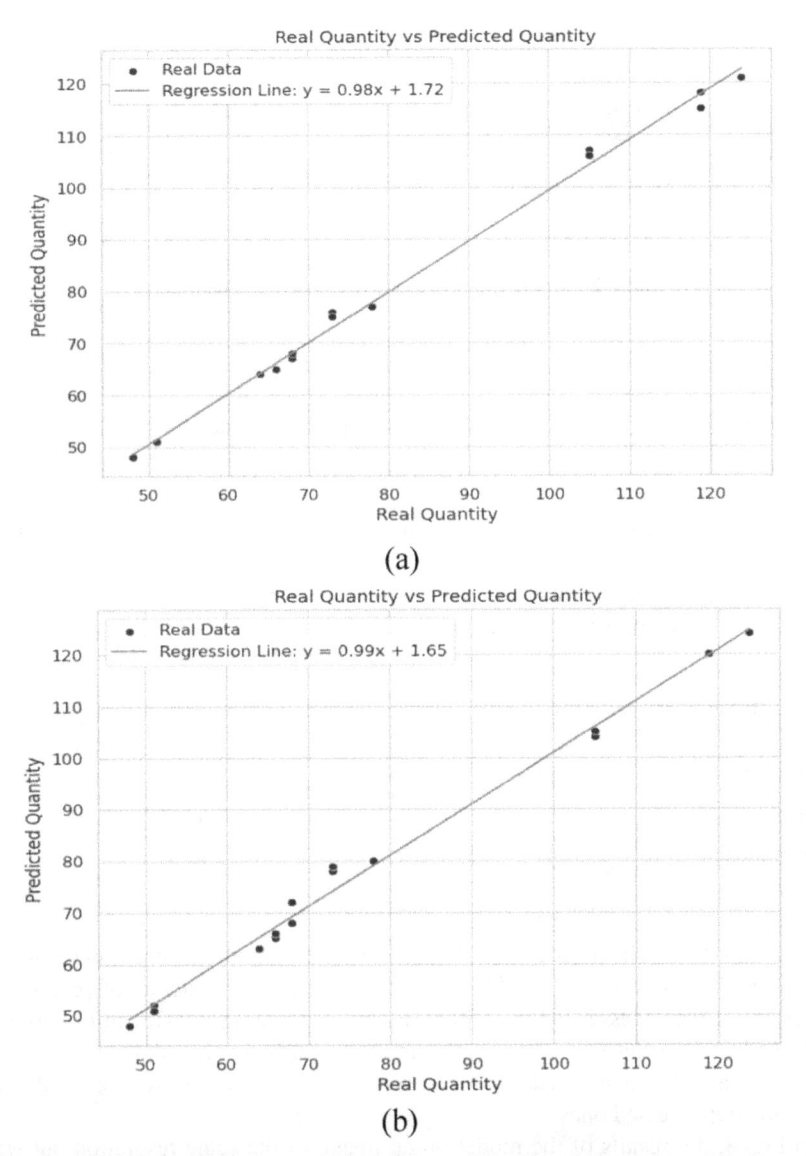

Fig. 7. Scatter plot relating quantities obtained by the algorithm and visual count, with the regression line showing the correspondence of both variables. (a) Resnet-50. (b) Resnet-101.

of the neural network itself as well as the counting performed by it. The major difference lies in the precision metrics and errors associated with the real count versus the count predicted by the model.

Despite the small differences in the mentioned metrics, it was found that ResNet-50 is the appropriate model for performing detections and segmentations of cell nuclei. This model not only yields the best results but is also lightweight and faster compared

Table 1. Model evaluation metrics.

Model	mAP50	Precision	Recall	F1-score	Effectiveness
Resnet-50	97.9	98.0	97.6	97.8	98.6
Resnet-101	97.7	95.9	97.4	96.7	98.0

Table 2. Model evaluation metrics based on quantities.

Model	MSE	RMSE	MAE	R^2	Max Error
Resnet-50	3.00	1.73	1.25	0.99	4
Resnet-101	5.44	2.33	1.44	0.99	4

to ResNet-101, which consumes more computational resources such as memory and processing time.

Regarding the results evaluated with ResNet-50 on the test set, we also analyzed the impact of including preprocessed images in the training and test sets.

Table 3. Resnet-50 evaluation metrics on original and preprocessed images.

Image type	Average Percentage Error	Effectiveness
Original	1.05	98.95
Preprocessed	1.64	98.36

Based on the results from Table 3, the model has higher effectiveness in recognizing original images, reaching nearly 99% (98.95%), whereas for the set of preprocessed images, the effectiveness is lower, at 98.36%. This provides us with the information that to achieve better results with our network, preprocessing before entering into the network is not necessary. It also indicates that the network has learned better from original images than from preprocessed ones.

In Fig. 8, the results of the model on an image of the same resolution but with a different number of nuclei are displayed. Figure 8.a has a real count of 51, where the model predicted 51, while Fig. 8.c has a real count of 105, and the model predicted 106.

Visually, it can be seen that the model developed in this work is considered important in aspects of detection, segmentation, and counting, which was already shown in the evaluation metrics where a mAP50 of 97.9% and an effectiveness of 98.6% between visual and predicted counts were obtained.

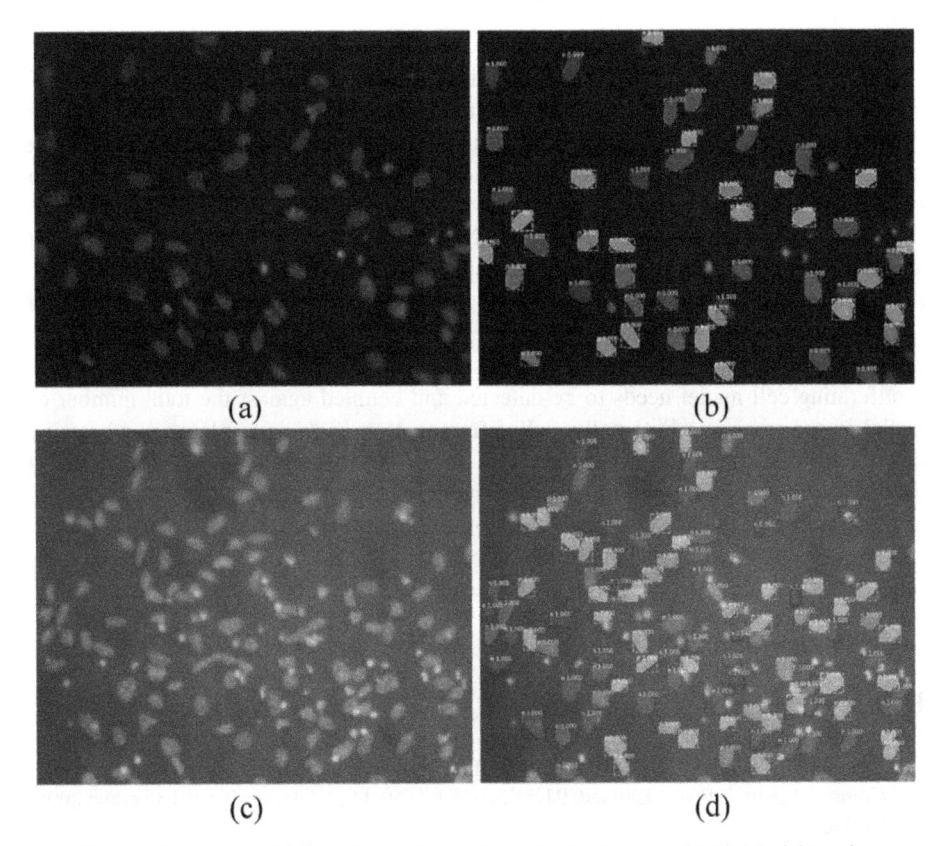

Fig. 8. Example of DAPI cell nuclei detections. (a) (c) Images. (b) (d) Model results.

4 Conclusion and Future Work

In this work, a Deep Learning-based artificial vision model was developed for the detection, segmentation, and counting of nuclei from Neural Progenitor Cells in fluorescence microscopy images. The proposed solution involved preprocessing and image enhancement stages, semi-automatic labeling of nuclei based on a Superpixel method, and training of an instance segmentation network for these nuclei.

The nuclei labeling was semi-automatic, images were preprocessed initially by applying filters and normalization to attenuate the noise and generate contrast between the objects of interest and the background. Then, Felzenszwalb Superpixel was applied to these images, whose segmentations required manual modification due to the presence of connected elements. This process allowed for the creation of a dataset of 44 images with a total of 3546 labels.

Preliminary tests using these labels and a Mask R-CNN neural network with Resnet-50 achieved a model capable of detecting cell nuclei with an accuracy rate of 97.7%. This model was used as an automatic object labeller in 22 new images, generating new labels. Despite the good performance of the model, certain false positives and false negatives had to be modified.

In the second stage, the neural network was trained by unifying the two datasets, resulting in a total of 66 images with 6392 labels. Tests were conducted on Mask R-CNN with Resnet-50 and Resnet-101 backbones to evaluate the impact of each backbone on this dataset. The model with the best results was obtained with Resnet-50, showing mAP50 metrics of 97.9% and an effectiveness of 98.6% between visual and predicted counts.

As future work, retraining of Mask R-CNN with new images verified by a geneticist specialist to improve its performance in result evaluation is planned, as well as adding different microscope sensors into the model. This work is part of a larger research project aimed at automatically analyzing the proliferation and differentiation behaviors of NPC cells. Future work will involve the analysis of cell proliferation, where the number of proliferating cell nuclei needs to be detected and counted against the total number of nuclei in the image. Additionally, cell differentiation analysis is planned, where the detection and distinction between each type of cell are required. In these cases, not only detections but also masks and counts of cell nuclei obtained by Mask R-CNN are crucial, as the presence of nuclei is essential to recognize different behaviors.

These results will contribute to the automation of laboratory processes, reduce the time spent on visual image analysis, generate decision support information, and provide tools for lab experts.

References

1. Allalou, A., Wählby, C.: BlobFinder, a tool for fluorescence microscopy image cytometry. Comput. Methods Progr. Biomed. **94**(1), 58–65 (2009). https://doi.org/10.1016/j.cmpb.2008.08.006
2. Harrer, S., Shah, P., Antony, B., Hu, J.: Artificial intelligence for clinical trial design. Trends Pharmacol. Sci. **40**(8), 577–591 (2019). https://doi.org/10.1016/j.tips.2019.05.005
3. Fuchs, T., Buhmann, J.: Computational pathology: challenges and promises for tissue analysis. Comput. Med. Imaging Graph. **35**(7–8), 515–530 (2011). https://doi.org/10.1016/j.compmedimag.2011.02.006
4. Xie, N., Li, X., Li, K., Yang, Y., Shen, H.: Statistical karyotype analysis using CNN and geometric optimization. IEEE Access **7**, 179445–179453 (2019). https://doi.org/10.1109/access.2019.2951723
5. Chadha, G., Srivastava, A., Singh, A., Gupta, R., Singla, D.: An automated method for counting red blood cells using image processing. Procedia Comput. Sci. **167**, 769–778 (2020). https://doi.org/10.1016/j.procs.2020.03.408
6. Haghofer, A., Dorl, S., Oszwald, A., Breuss, J., Jacak, J., Winkler, S.: Evolutionary optimization of image processing for cell detection in microscopy images. Soft Comput. (2020). https://doi.org/10.1007/s00500-020-05033-0
7. Kan, A.: Machine learning applications in cell image analysis. Immunol. Cell Biol. **95**(6), 525–530 (2017). https://doi.org/10.1038/icb.2017.16
8. Xing, F., Xie, Y., Su, H., Liu, F., Yang, L.: Deep learning in microscopy image analysis: a survey. IEEE Trans. Neural Netw. Learn. Syst. **29**(10), 4550–4568 (2018). https://doi.org/10.1109/tnnls.2017.2766168
9. Moen, E., Bannon, D., Kudo, T., Graf, W., Covert, M., Van Valen, D.: Deep learning for cellular image analysis. Nat. Methods **16**(12), 1233–1246 (2019). https://doi.org/10.1038/s41592-019-0403-1

10. Gupta, A., et al.: Deep learning in image cytometry: a review. Cytometry Part A **95**(4), 366–380 (2018). https://doi.org/10.1002/cyto.a.23701
11. Tsai, H., Gajda, J., Sloan, T., Rares, A., Shen, A.: Usiigaci: instance-aware cell tracking in stain-free phase contrast microscopy enabled by machine learning. SoftwareX **9**, 230–237 (2019). https://doi.org/10.1016/j.softx.2019.02.007
12. Mata, G., et al.: Automated neuron detection in high-content fluorescence microscopy images using machine learning. Neuroinformatics **17**(2), 253–269 (2018). https://doi.org/10.1007/s12021-018-9399-4
13. Carneiro, G., Zheng, Y., Xing, F., Yang, L.: Review of deep learning methods in mammography, cardiovascular, and microscopy image analysis. In: Lu, L., Zheng, Y., Carneiro, G., Yang, L. (eds.) Deep Learning and Convolutional Neural Networks for Medical Image Computing. Advances in Computer Vision and Pattern Recognition, pp.11–32. Springer, Cham (2017). https://doi.org/10.1007/978-3-319-42999-1_2
14. Klauschen, F., et al.: Scoring of tumor-infiltrating lymphocytes: From visual estimation to machine learning. Semin. Cancer Biol. **52**, 151–157 (2018). https://doi.org/10.1016/j.semcancer.2018.07.001
15. Sommer, C., Gerlich, D.: Machine learning in cell biology – teaching computers to recognize phenotypes. J. Cell Sci. **126**(24), 5529–5539 (2013). https://doi.org/10.1242/jcs.123604
16. Kaggle.com. 2018 Data Science Bowl—Kaggle (2022). https://www.kaggle.com/c/data-science-bowl-2018. Accessed 11 Apr 2024
17. Johnson, J.W.: Adapting Mask-RCNN for Automatic Nucleus Segmentation. Advances in Intelligent Systems and Computing (2020). https://doi.org/10.1007/978-3-030-17798-0
18. Toro, C.O.: Algoritmos de segmentación semántica para anotación de imágenes. https://doi.org/10.20868/upm.thesis.55407
19. Li, H., Wei, D., Cao, S., Ma, K., Wang, L., Zheng, Y.: Superpixel-Guided Label Softening for Medical Image Segmentation. https://doi.org/10.48550/arXiv.2007.08897
20. Bechar, M., Settouti, N., Barra, V., Chikh, M.: Semi-supervised superpixel classification for medical images segmentation: application to detection of glaucoma disease. Multidimens. Syst. Signal Process. **29**(3), 979–998 (2017). https://doi.org/10.1007/s11045-017-0483-y
21. Kapuscinski, J.: Dapi: A DNA-specific fluorescent probe. Biotech. Histochem. **70**(5), 220–233 (1995). https://doi.org/10.3109/10520299509108199
22. Felzenszwalb, P., Huttenlocher, D.: Efficient graph-based image segmentation. Int. J. Comput. Vis. **59**(2), 167–181 (2004). https://doi.org/10.1023/b:visi.0000022288.19776.77
23. He, K., Gkioxari, G., Dollár, P., Girshick, R.: Mask R-CNN. https://doi.org/10.48550/arXiv.1703.06870
24. He, K., Zhang, X., Ren, S., Sun, J.: Deep residual learning for image recognition. https://doi.org/10.48550/arXiv.1512.03385
25. Lin, T.-Y., et al.: Microsoft Coco: Common Objects in Context. https://doi.org/10.48550/arXiv.1405.0312

Object Recognition Models for Indoor Users' Location

Franco M. Borrelli[1,2(✉)] and Cecilia Challiol[1,3]

[1] LIFIA, Facultad de Informática, UNLP, La Plata, Buenos Aires, Argentina
{fborrelli,ceciliac}@lifia.info.unlp.edu.ar
[2] UNLP Master's Scholarship, La Plata, Argentina
[3] CONICET, Buenos Aires, Argentina

Abstract. Despite technological advances, precise positioning within buildings remains a considerable challenge. In this context, the present paper explores the research of user location in indoor spaces, embracing object recognition models executed directly on mobile devices. Our proposal is based on designing a generic solution architecture adaptable to any physical environment, enabling the definition and usage of relevant generic objects within the environment to determine the users' current location. This proposal uses Computer Vision, employing object recognition models for positioning. This kind of indoor positioning benefits from the growth of smartphones' functionalities and capabilities, thus avoiding the need to install additional infrastructures in physical spaces. A specific implementation of this architecture for *React Native* is presented, using the *TensorFlow* platform to support object recognition. This implementation allows demonstrating how this positioning works through concrete use cases. In addition, some lessons learned are discussed, which we hope will contribute to this topic.

Keywords: Object Recognition Models · Indoor Location · User Location · Lightweight Networks

1 Introduction

The general adoption of mobile devices, coupled with technological advancements in recent years, has facilitated the growth of a diverse range of mobile applications, particularly those context-aware [1, 2]. These applications use, for instance, environmental data surrounding the user to provide different kinds of services or information. Location is a relevant context in these applications, which can be obtained using different technologies such as GPS, Wi-Fi, or Bluetooth [3]. While this context has been extensively utilised in outdoor spaces thanks to the advantages of GPS, resolving the users' location in indoor spaces still poses a challenge [3].

Although several solutions have been proposed for indoor locations, a scalable and effective solution that works universally has not yet been achieved. In [4] and [5], the use of beacons as a location-sensing mechanism is explored, while in [6] and [7], Wi-Fi network fingerprinting is employed for user location. Besides, these technologies require

© The Author(s), under exclusive license to Springer Nature Switzerland AG 2025
M. Naiouf et al. (Eds.): JCC-BD&ET 2024, CCIS 2189, pp. 30–44, 2025.
https://doi.org/10.1007/978-3-031-70807-7_3

installing additional infrastructure to work in the environment. For instance, beacons require the installation of numerous battery-operated devices that transmit Bluetooth signals, demanding regular maintenance. Something similar occurs with Wi-Fi fingerprinting, which implies the installation of at least three Wi-Fi antenna networks. Therefore, deploying and maintaining such infrastructures can be expensive, which could be a limitation.

On the other hand, the approach known as *vision-based indoor positioning* [3] has emerged in recent years. This approach proposes using object recognition models to locate users in indoor spaces. Although promising ad-hoc solutions have been developed using this approach [3], generalised implementations are still lacking. This is part of the motivation of this paper.

Object recognition (and detection) aims to enable computer systems to identify or classify objects in images and videos in an automated and precise manner [8]. In the traditional vision, the execution of recognition models is linked to data centers and large clusters of machines with powerful GPUs. However, this alternative can be costly and requires transferring all mobile device data and sending them through a network connection, which can be time-consuming.

Technological advancements allow nowadays to perform object recognition locally on mobile devices [9] using *lightweight networks* [8, 10]. This is a new area of research, and its viability for particular uses still needs to be evaluated [8]. Furthermore, as far as we know, *lightweight networks* are little explored for positioning in indoor spaces. This paper contributes in this direction.

Another aspect to mention is that currently, the implementation of object recognition and detection models has been significantly simplified with the development of platforms such as *TensorFlow*[1], which offers a wide range of pre-trained models that are ready to use. Additionally, these platforms often provide facilities for training new models using techniques such as transfer learning [11].

Considering the above mentioned, this paper presents a generic solution architecture for positioning users in indoor spaces using object recognition models that run locally on mobile devices (via *lightweight networks*). This proposal allows for the definition and utilisation of any relevant generic object within the environment to position users, and this positioning can be embedded in different kinds of applications. It is essential to highlight that this solution aims to leverage the smartphones' capabilities and avoid the need to install additional infrastructure.

Furthermore, this paper presents a specific implementation of the proposed architecture to demonstrate, through use cases, how this positioning works. In particular, a library was implemented in *React Native*[2] combined with the *TensorFlow* library for *Javascript* to support object recognition. This paper also discusses some lessons learned, which we hope will contribute to this field.

This paper is structured as follows. Section 2 describes the state of the art regarding object recognition and its applicability for positioning, as well as some details of *lightweight networks*. Section 3 presents a generalised solution architecture for indoor positioning using object recognition. A specific instantiation of this architecture is

---
[1] Tensorflow. https://tensorflow.org/, last accessed 2024/03/15.
[2] React Native. https://reactnative.dev/, last accessed 2024/03/15.

described in Sect. 4. Section 5 illustrates through use cases how this instantiation works. Finally, Sect. 6 presents some conclusions and future work.

2 Review of Literature

Object recognition and detection have experienced remarkable evolution in the last two decades [8, 10]. This section explicitly addresses two concepts: *vision-based indoor positioning* and *lightweight networks*, which form the foundation of the research presented in this paper.

2.1 Vision-Based Indoor Positioning

An interesting use of object recognition explored in recent years is *vision-based indoor positioning* [3], which seeks to employ object detection to guide and locate the user within indoor spaces. The concept of *vision-based indoor positioning* was first coined by Microsoft [12] in 1998. At that time, Microsoft's Vision Technology group analysed how technology could facilitate life in the future, and one of the proposed fundamental technologies was locating people in the home using various video surveillance techniques. Since then, this concept has evolved to three possible implementation options [3], which vary in how information is stored and processed. These are detailed below:

- The first one is known as *'object-based reference'*, which focuses on detecting static objects in images and then searching for correspondences of these objects with a database of buildings (often represented as 3D models) containing information about the objects' location within them. This approach can be executed through various methods; for example, in [3], a *'control point'* method is used where each object of interest in space is composed of a series of these points, representing physical characteristics with associated precise coordinates. Subsequently, an algorithm compares the distances between the coordinates detected by the camera with those recorded in a database, allowing the calculation of the user's distance from these objects. Although the results obtained in [3] are promising, it is essential to note a significant limitation: the used model requires training with specific environment data to identify control points correctly. This requirement limits the use of this solution in different environments. This motivates our proposed generic architecture in this paper, which is adaptable to various physical spaces where databases may contain generic objects that could be present in different environments.
- The second alternative is known as *'reference from images'*, which proposes comparing previously taken images of specific routes within the building with the current view of the device's camera. This approach requires an initial data collection stage where images covering relevant locations of the environment should be taken. Then, these images must be labelled by establishing connections between visual elements and specific locations within the building. Subsequently, on real-time execution, the device's camera's current view is compared with previously captured image sequences. The system uses algorithms to identify correspondences between the current scene and the stored images. The primary obstacle of this approach resides in achieving real-time positioning capability because image correspondence involves an exceptionally

high computational load. Additionally, the *'reference from images'* option is highly coupled to specific environments, making achieving a generalisable solution to any physical space more complex.
- Finally, the use of *'reference from deployed coded targets'* proposes using objects such as barcodes, QR codes, and dot patterns. For this, it is necessary to generate the markers, link them with relevant information about the place, and physically place them in their corresponding location. This option presents the inherent challenges in using QR codes, such as their deterioration. Additionally, these codes must be generated and placed in relevant locations for each physical space, making this kind of approach ad-hoc for each environment.

As mentioned before, the *'reference from images'* option is highly joined to specific environments because it requires the development of models capable of recognising specific buildings' features, making it less versatile and too complex to propose a scalable solution. The case of the *'reference from deployed coded targets'* presents previously mentioned challenges, such as being ad-hoc for each environment. For these reasons, this paper focuses on exploring the first alternative, *'object-based reference'*, proposing a flexible and adaptable architecture for various physical spaces. In this architecture, the objects in the databases are generic and could, therefore, be present in different environments.

2.2 Lightweight Networks

In recent years (from 2017 onwards), a new branch of research has emerged within the Computer Vision and Deep Learning areas, known as *lightweight networks* [8, 10]. These investigations aim to design small and efficient networks for environments with limited resources, such as mobile and IoT devices. Generally, these networks use techniques such as pruning, quantisation, and hashing to improve model efficiency. A common characteristic of all these networks is that they sacrifice accuracy for increased speed in inference. As mentioned in [8], several architectures make use of *lightweight networks*, such as *MobileNet*, *ShuffleNet*, and *Once-For-All (OFA)*, all of them present two interesting features:

- They can adequately recognise objects in real-time on medium to high-end devices, ensuring their utility in a wide variety of mobile devices.
- Additionally, they run locally on the device, eliminating the need for an internet connection and optimising both response speed and user experience.

The *lightweight networks* are promising but still face some outstanding challenges that should be resolved [8–10, 13]. For example, one of the main issues is that a solution is still needed to address detection on video streams for these devices [10]. Furthermore, the speed gap between a machine and human eyes remains large, especially for detecting small objects [10]. Moreover, *lightweight networks* are still few explored for user positioning in indoor spaces; thus, this paper aims to contribute to this area.

3 Proposed Architecture for Object-Based Positioning

This section proposes a generic architecture for any physical space, allowing the creation and use of relevant generic objects within the environment to users' location. According to that, two key aspects were considered when this architecture was designed, such as:

- A mechanism is required to recognise objects present in the physical space. This task will be addressed through object recognition and detection models capable of recognising generic objects in the environment.
- It is essential to have additional information so that, from the detected objects, the user can be contextualised and located in a specific physical space.

To ensure that both aspects mentioned above work to locate users effectively, two stages are needed: one focused on defining and creating these relevant positions, and the other centred on using this information to infer the user's current location. Below, each of these stages is detailed:

- *Creation*: The architecture needs to provide a mechanism for users to register in-situ data in the contextual object database. This database which will store detailed information about each detected (and relevant) object, linking each one with its respective spatial location and other relevant data. This location could be taken with any positioning mechanism. The solution must be flexible and extensible to enrich the data that could be loaded in the future.
- *Usage*: Once the contextual object database is built, the available information can be used to locate a user. When the user points the device's camera at an object, it is determined whether the object is registered in the contextual database and whether the user is *'near'* the registered location for that object. Thus, in case of success, an application (using this kind of positioning) can, for example, provide relevant information or specific services associated with that location and recognised object.

Figure 1 depicts the proposed architecture. Different applications can embed the *Object-based Positioning* module to load the contextual database (*creation*) or access previously loaded data (*usage*) and thus use this positioning mechanism to locate the user. Additionally, applications could use both functions (*creation* and *usage*). It is important to mention that the contextual database could be local or not to the mobile device, depending on the requirements and complexity of the implementation.

Furthermore, Fig. 1 shows that the *Object-based Positioning* knows and observes two modules: one for object recognition and another for positioning. *Object-based Positioning* observes the modules because it constantly *'listens'* for any changes (whether a new detected object or a new location). The *Object-based Positioning* can also communicate directly with these modules (through the *'knows'* relationship) to request information, which usually occurs in response to the user's action. Below are more details about these two modules.

Object Recognition Models for Indoor Users' Location 35

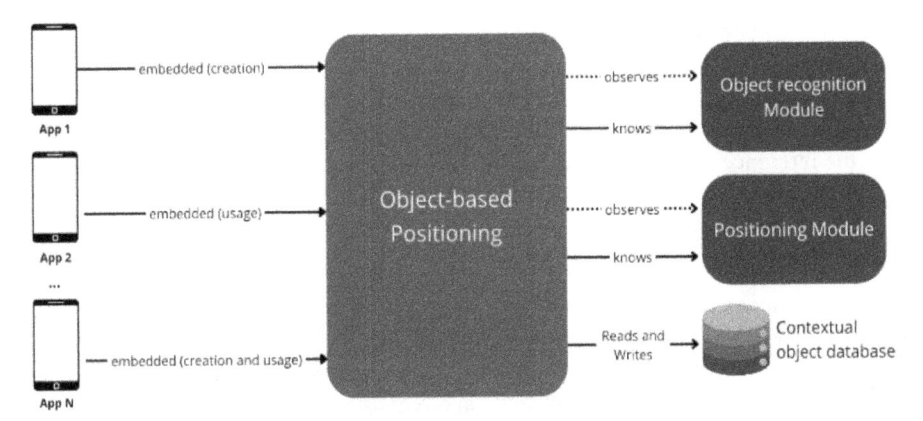

Fig. 1. Proposed Architecture for *Object-based Positioning*.

The *Object Recognition* module recognises objects in the environment through the device's camera. It can know several models, allowing it to be extensible, but the module only uses one to perform the recognition at a given time. Figure 2 depicts the flow involved in this module, which implies three specific steps.

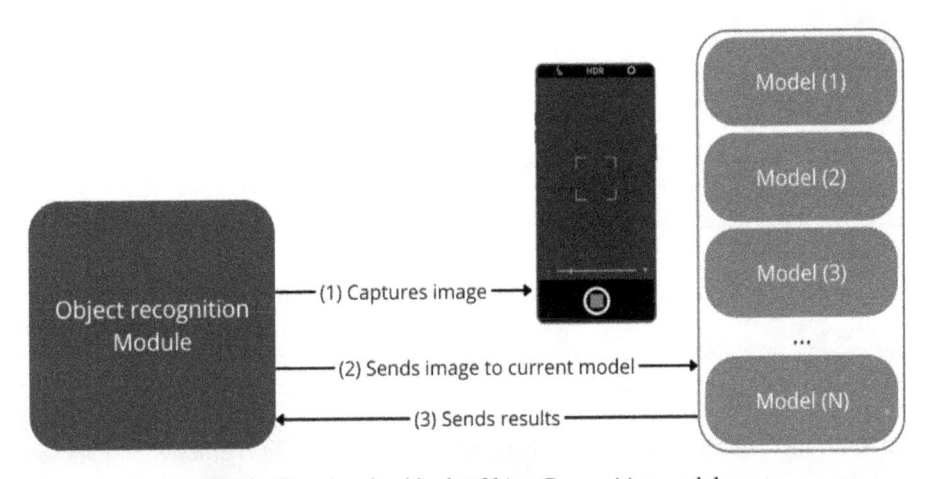

Fig. 2. Flow involved in the *Object Recognition* module.

To provide a clear understanding of the three steps of the *Object Recognition* module depicted in Fig. 2, they are detailed below, using the same enumeration as in the figure:

1. The current image is obtained, which involves establishing communication with the device's camera to capture the view at that moment.
2. The captured image is sent to the current object recognition or detection model. Depending on the chosen model, some pre-processing may be necessary to adapt the image to the format expected by the model. Additionally, the *Object Recognition* module allows the configuration of the model to be used and the selection of the

objects of interest during the inference process, which is especially relevant when employing *Object-based Positioning*.

3. Finally, the corresponding result is received once the model inference is performed. It is important to highlight that the format and structure of the response obtained in the inference may vary from one model to another. Therefore, before returning a final response, the *Object Recognition* module standardises the response to a standard interface. This facilitates communication with different object recognition and detection models.

On the other hand, the *Positioning module* aims to obtain information about the user's location using various positioning sensing mechanisms, as shown in Fig. 3. This module is designed to constantly monitor changes in the selected sensing mechanism, allowing it to detect events such as user movement and request updated location information as needed. The module provides a standardisation layer regarding the response provided by each sensing mechanism to ensure coherence and uniformity of location data. The chosen sensing mechanism may depend on the required accuracy, infrastructure availability, and environmental limitations.

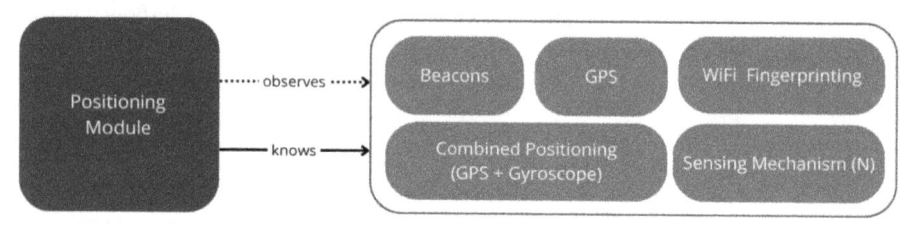

Fig. 3. Representation of the *Positioning* module.

4 An Implementation of the Object-Based Positioning

This section presents a concrete implementation of the *Object-based Positioning* module from the architecture proposed in Sect. 3. This module was implemented as a library[3] for the *React Native* framework (which can create cross-platform applications for Android and iOS with the same source code). Note that the *Object-based Positioning* library can be embedded by any application, as presented in Fig. 1.

React Native offers various modules and libraries that facilitate integration with native phone functionalities. These include access to the device's camera through the *React Native Camera* library and the execution of object recognition models with the specific *Tensorflow* library for *React Native*. In particular, the library has been implemented to allow the use of two models based on *MobileNet*. The first one is a model for recognising doors (trained by one of the authors of this paper), and the other is a pre-trained *TensorFlow* model that can recognise various general objects.

---

[3] The project's source code is available at the following GitHub repository: https://github.com/francoborrelli/object-based-positioning-library. This repository contains all the documentation and files needed to understand and utilize the *Object-based Positioning* library.

Notably, the *Positioning module* was chosen to explore the combination of GPS and Gyroscope. This does not require the installation or presence of additional infrastructure in the physical environment; that is, everything is solved from the mobile device, which simplifies the use of this architecture in any physical location. This facilitates obtaining a generic solution that could be used in any environment. *React Native* has the *React Native Geolocation* and *React Native Sensors* libraries, which allow access to the device's GPS and sensors (such as the Gyroscope), respectively.

Firestore was chosen for the contextual database. It is a NoSQL cloud database from *Google Firebase*[4] that offers efficiency, scalability, and an SDK for multiple platforms. Thus, a *Firestore* database is used to store locations that involve data collected from the object recognition and positioning modules. Figure 4 shows the technologies used for the implemented *Object-based Positioning*.

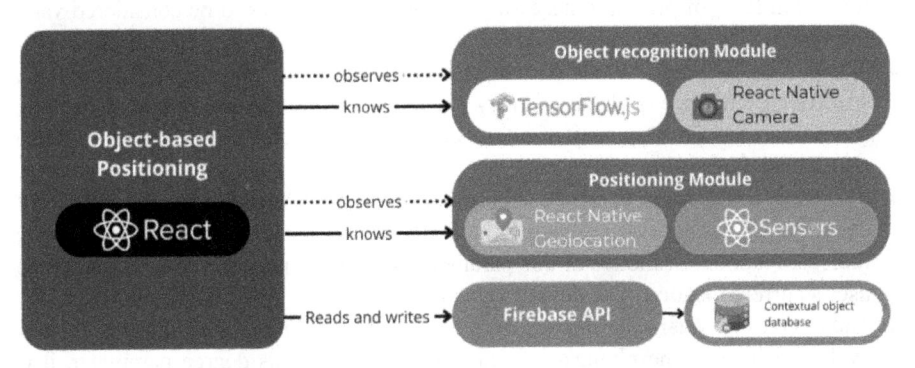

Fig. 4. Technologies used in the *Object-based Positioning* library.

5 Practical Application of the Object-Based Positioning Library

The topic of positioning has been previously explored by the authors of this paper. In particular, in earlier works, positioning was investigated using Wi-Fi network fingerprinting to locate the user in in-situ co-design authoring tools [6, 7]. Based on this, it was decided to continue with this line of research, but in this paper, the implemented *Object-based Positioning* library is used as the positioning mechanism. Below, we detail how the *Object-based Positioning* library was embedded in an in-situ co-design authoring tool, and then we exemplify its usage, emphasising the lessons learned.

5.1 Object-Based Positioning Library Embedded in an In-situ Co-design Tool

The *Object-based Positioning* library was embedded into an in-situ co-design tool, which used the two functionalities provided by this kind of positioning, such as *'creation'* and *'usage'* (described in Sect. 3). This tool focuses, in particular, on co-designing (creating) *Positioned Information* using object-based positioning to subsequently provide (use) this

---
[4] Firebase. https://firebase.google.com, last accessed 2024/03/15.

information when a user is in that location. For more details on the functioning of in-situ co-design tools, [6] and [7] can be consulted.

Some configurable parameters related to *Object-based Positioning* were added to the tool to facilitate an easy way to change these values through an interface. This allows the creation of different scenarios to explore how the library behaves with different values. These configurable parameters are:

- *Recognition or detection model to use at one point in time*. It can choose from the two models based on *MobileNet* (one focuses only on doors, and the other is oriented to general objects), as mentioned in Sect. 4.
- *Classes of interest*. Depending on the selected model, it can choose among the available objects, which the object-based positioning will consider.
- *The minimum precision adjustment in inferences*. This parameter allows setting the minimum level of precision necessary for the model outputs to be considered valid inferences. The default value is 20%.
- *The number of frames*. This parameter controls how recognition models operate, specifying inference on 1 out of every N frames from the phone camera. Lowering it will increase the frequency of inferences, though resource consumption will also rise. The default value is 100 frames.
- *The maximum distance in meters to consider an object is 'near' to the user*. An object detection is only considered valid if the user is at a distance less than or equal to the indicated amount of meters of this parameter. GPS information is used to obtain the user's current location. The default value is 1 m.
- *The degrees to determine that an object is 'in the viewing angle' of the user*. It is evaluated if, with the phone's current orientation plus this degree parameter, it is possible to *'see'* a detected object according to the viewing angle registered for this object. Gyroscope information is used to obtain the user's current orientation. The default value is 70°.

5.2 Usage of the Object-Based Positioning Library - Single Object Class

This prototypical experience was designed to assess the functionality of *Object-based Positioning* when registering *Positioned Information* using multiple objects of the same type. To achieve this, the parameter *'Recognition or detection model to use at one point in time'* was configured with the door recognition model presented in Sect. 4. For this experience, *Positioned Information* was created using the domain of a hypothetical conference that could take place between classrooms 1–2 to 1–4 on the first floor of the Faculty of Informatics at UNLP, assigning a specific track (topic) to each classroom. Then, it could use this *Positioned Information* (for each of these tracks) to provide them to the users when they were in those locations.

Specific aspects were established to observe how the library behaved in response to nearby doors (classrooms 1–2 and 1–3 of the faculty), such as variation in viewing angles when registering information, changes in user orientation, and walking speed. It should be noted that some of these aspects to observe emerged from our previous knowledge using other location mechanisms.

During the *'creation'* phase, two participants were asked to load *Positioned Information* for the different classrooms into two distinct Workspaces[5]. In one, the information was loaded facing the doors, while in the second, different angles were used. Upon observing the recorded information for both Workspaces, a positioning discrepancy between the devices used (Xiaomi 12 and Samsung S22) was noticed. To investigate this issue further, a direct comparison was made between the latitude and longitude readings of both devices, as shown in Fig. 5. This test revealed a 6-meter difference between the locations recorded by the two devices, even though they were physically next to each other.

Several factors were identified that could have contributed to the discrepancies in GPS values (shown in Fig. 5), such as antenna positioning, the type of GPS technology used, GPS calibration, the devices' power-saving mode, among others. Solving this is beyond the scope of this paper.

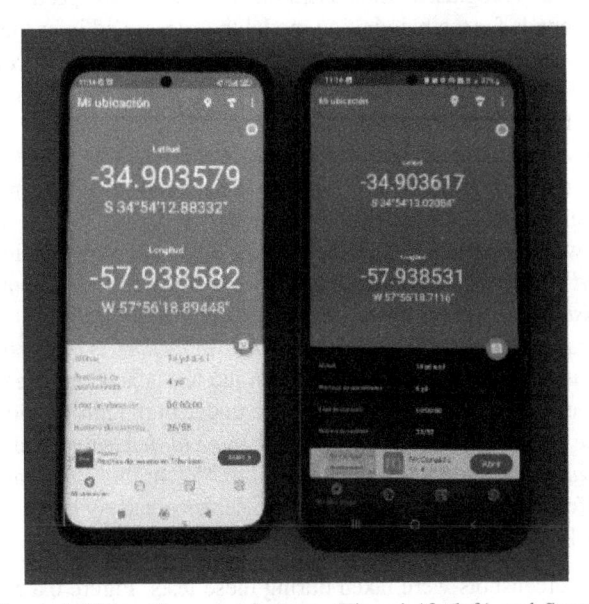

Fig. 5. Comparison of GPS performance between Xiaomi 12 (left) and Samsung S22 (right) devices.

Considering these disparities with the GPS, it was decided that a single participant (the one who had the minor discrepancy in positioning) would be responsible for loading all the *Positioned Information* since the focus was on testing the library rather than *'improving'* GPS performance.

When the selected participant finished loading all the *Positioned Information*, both participants took part in a *'usage'* phase. During this last stage, some interesting points were detected that are worth mentioning:

---

[5] Each Workspace contains all the information co-designed within an application. Due to the nature of this tool (for in-situ co-design), each Workspace is linked to a building or physical location.

- With the default settings, which include a *'The maximum distance in meters to consider an object is near to the user'* of 1 m and a viewing angle aperture of 70°, participants were asked to slowly approach the classroom doors to see if they could access the information. Only the participant who had loaded the information could access it by doing this. The other participant experienced difficulties, which were partly expected given the discrepancies detected with the GPS. This was done in both Workspaces, where the same outcome was experienced.
- Based on the previous findings, the parameter *'The maximum distance in meters to consider an object is near to the user'* was changed from 1 to 3 m. With this adjustment, both participants could access the information from all classrooms correctly in both Workspaces.
- The other test involved participants being asked to approach the classroom doors at a much faster pace. The result was the same in both Workspaces. It was identified that objects cannot be recognised when users move at very high speeds, as it requires at least a few seconds for the recognition model to process and infer the object.
- As in all previous tests, participants could access the correct information for each classroom. It was desired to determine at what distance the library began to mix information, for example, between the content of classrooms 1–2 and 1–3, which are physically very close to each other. Only when reaching a vast *'The maximum distance in meters to consider an object is near to the user'* setting (10 m) did a problem of information confusion start to appear.
- An analysis was conducted on how the viewing angle aperture affected the detection of *Positioned Information*. Two parameters were adjusted for this purpose: the *'The maximum distance in meters to consider an object is near to the user'* was set to 3 m in both Workspaces and the *'The degrees to determine that an object is in the viewing angle of the user'* was reduced to 30°; when the tests carry on revealed that with a 30-degree aperture, it was no longer possible to detect the information.
- Based on the previous findings, the *'The degrees to determine that an object is in the viewing angle of the user'* was increased to 45°. In this case, participants could detect the information if they positioned themselves precisely in the same place with the same angle used to register the information.

Images and screenshots were taken during these tests. Figure 6.a reflects the *'creation'* stage, where it can be observed how the tool can recognise different elements of the environment (in this case, doors). The user can choose one of the objects identified by the model and load *Positional Information* related to it. In Fig. 6.a, the loading form can also be seen along with some positioning data (such as the object detected and latitude and longitude information provided by the GPS; the Gyroscope angle is not seen because it is below the screen but is considered too). Figure 6.b presents an example of *'usage'*, where the user approaches the registered door and accesses its content.

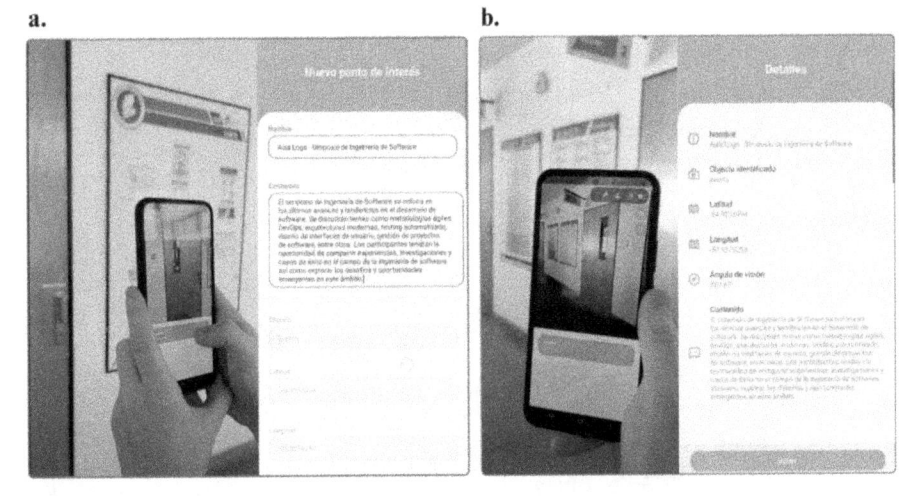

Object Recognition Models for Indoor Users' Location 41

Fig. 6. Images and screenshots taken during the *Object-Based Positioning* library testing.

5.3 Usage of the Object-Based Positioning Library - Multiple Object Classes

This experience aimed to develop a hypothetical application for children visiting an informatics exposition. To simulate this, a booth was set up with various technological objects to offer interesting facts about them (see Fig. 7.a). The objective was to analyse the behaviour of the *Object-based Positioning* library when different objects are used in a reduced space. For this experience, two participants were also involved: the first with a high-end phone (Samsung S22) and the second with a mid-range phone (Xiaomi Mi A3).

Before starting, a Workspace was defined, and the parameter *'Recognition or detection model to use at one point in time'* was set with the pre-trained *MobileNet* model capable of recognising various general objects. This model can recognise laptops, mice, cellphones, hard drives, etc. Only the objects in Fig. 7.a were selected for the parameter *'Classes of interest'*. This decision was made to prevent unwanted objects from being detected with the tool.

During the *'creation'* phase, one participant loaded the *Positioned Information* with the tool, considering the issue detected in Sect. 5.2 with the GPS accuracy. Although the GPS accuracy was not precise even with the same device, it remained within acceptable limits; this margin of error is shown in Figs. 7.a and 7.b. Figure 7.a represents how the objects were actually distributed, while Fig. 7.b shows how the *Positioned Information* was recorded in the tool. Additionally, in Figs. 7.c and 7.d, two examples of detections with the tool are presented, showing the identified object's name and the model's corresponding precision percentage.

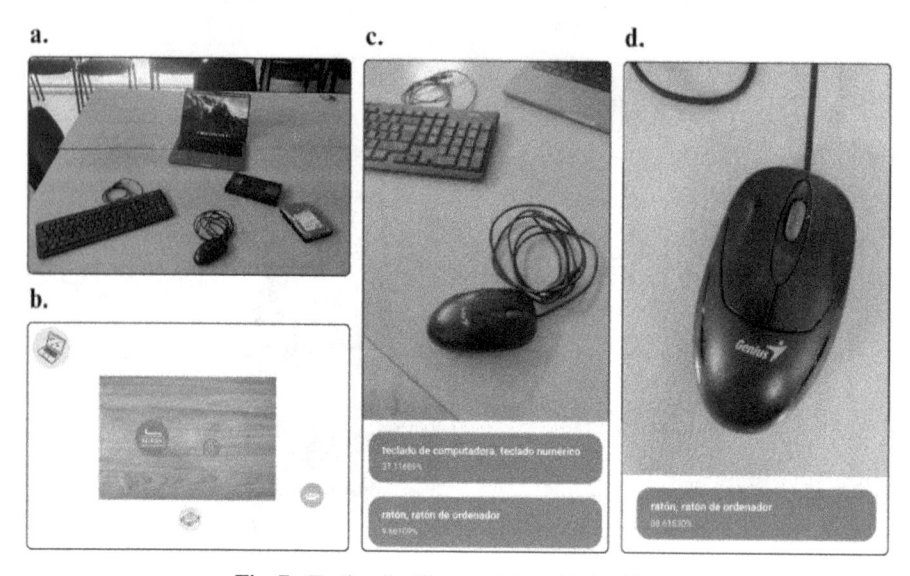

Fig. 7. Testing the library with multiple objects.

During the *'usage'* phase, both participants were asked to try accessing the *Positioned Information* while facing the stand. Several tests were conducted, and some interesting points were detected that are worth mentioning:

- Both participants encountered difficulties with the default settings (defined in Sect. 5.1). This was due to the previously identified GPS issues, which prevented them from consistently accessing the *Positioned Information*. Specifically, the participant who defined the information could only recognise some objects, while the others failed to recognise any of the objects.
- Based on the previous results, the parameter *'The maximum distance in meters to consider an object is near to the user'* was changed from 1 to 3 m. Both participants were able to access all the information with this setting.

These results, combined with the experience of Sect. 5.2, revealed that *'The maximum distance in meters to consider an object is near to the user'* of 3 m is the most optimal. Furthermore, it was interesting to test how the recognition model performed with different configurations of *'The minimum precision adjustment in inferences'* and *'The number of frames'*. Below is a description of what was discovered:

- When using the *'The minimum precision adjustment in inferences'* configuration with its default value (20%), recognition of all objects was achieved, albeit with a delay of approximately 5 s.
- However, by reducing *'The minimum precision adjustment in inferences'* to 10%, recognition occurred almost instantly, with precision levels close to 17%.
- Upon further analysis of the *'The minimum precision adjustment in inferences'*, it was observed that the accuracy of the model gradually increased over time, always starting with a very low value (see Fig. 7.c) and gradually increasing within seconds to much higher values (between 40% and 80% in most cases), as shown in Fig. 7.d.

- In addition, it was found through another test that the time window during which the model achieves high precision percentages can be significantly reduced by modifying *'The number of frames'* configuration. This configuration was changed from its default value (100 frames) to a much lower one (10 frames) in the conducted test. This caused the tool to instantly generate results with higher precision when using the Samsung S22 device. However, this modification caused the application to experience inevitable hitches with the Xiaomi Mi A3, as this phone is a mid-range device with 4 GB of memory.

6 Conclusions and Future Works

The implemented tool, which embeds our *Object-based Positioning* library, has demonstrated the practical viability of the proposed general architecture for indoor positioning using object recognition models. Although preliminary in nature, the conducted tests have provided valuable insights into the potential adoption of this positioning approach. During these tests, it was identified that the implemented *Object-based Positioning* solution works for recognising different objects and detecting objects of the same type in limited spaces. Furthermore, the tests confirmed that the solution can operate effectively on high-end and mid-range devices.

It should be noted that depending on the space's features, the objects within it, and the devices being used, the settings of the implemented library may need to be adjusted to achieve optimal performance. For instance, it was found that adjusting with low values of the *'The number of frames'* configuration on a high-end device resulted in significantly better performance but led to operational issues on mid-range devices.

The Gyroscope's functionality was tested, and it was found that the phone's Gyroscope was effective in disambiguating nearby information. In particular, a *'The degrees to determine that an object is in the viewing angle of the user'* configuration of 70° yielded satisfactory results. Note that, with these few tests, we cannot reach conclusive results, so more exploration is needed to confirm this preliminary discovery.

Concerning GPS functionality, it was noted that the performance can exhibit variations across different devices even when they are physically situated in the same exact location. Although GPS issues were detected, the proposed solution is a viable alternative. Related to this, the parameter *'The maximum distance in meters to consider an object is near to the user'* should be configured with 3 m to get optimal results.

As a future direction, it would be valuable to investigate the implementation of the *Object-based Positioning* solution with other technologies such as Wi-Fi positioning, Bluetooth, or proximity sensors, considering the limitations identified with GPS. This could hold promise for enhancing the accuracy and reliability of indoor positioning, but other issues could emerge. For example, the feasibility of this future implementation will necessitate an assessment of additional infrastructure requirements, including their cost-effectiveness and scalability across various physical spaces.

References

1. Alegre, U., Augusto, J.C., Clark, T.: Engineering context-aware systems and applications: a survey. J. Syst. Softw. **117**, 55–83 (2016). https://doi.org/10.1016/j.jss.2016.02.010

2. Alegre-Ibarra, U., Augusto, J.C., Evans, C.: Perspectives on engineering more usable context-aware systems. J. Ambient. Intell. Humaniz. Comput. **9**(5), 1593–1609 (2018). https://doi.org/10.1007/s12652-018-0863-7
3. Xiao, A., Chen, R., Li, D., Chen, Y., Wu, D.: An indoor positioning system based on static objects in large indoor scenes by using smartphone cameras. Sensors (Basel Switz.) **18**(7), 2229 (2018). https://doi.org/10.3390/s18072229
4. Kaulich, T., Heine, T., Kirsch, A.: Indoor localisation with beacons for a user-friendly mobile tour guide. KI-Künstliche Intelligenz **31**(3), 239–248 (2017). https://doi.org/10.1007/s13218-017-0496-6
5. Borrelli, F.M., et al.: Desarrollo multiplataforma de Aplicaciones Móviles combinadas con el uso de Beacons. In: XXIV Congreso Argentino de Ciencias de la Computación (CACIC 2018), pp. 847–856. RedUNCI, Tandil (2018)
6. Challiol, C., et al.: Design thinking's resources for in-situ co-design of mobile games. In: 2019 International Conference on Information Systems and Computer Science (INCISCO 2019), pp. 339–345. IEEE, Quito (2019). https://doi.org/10.1109/INCISCOS49368.2019.00060
7. Challiol, C., et al.: Co-diseño in-situ de Juegos Móviles usando un abordaje con recursos de Design Thinking. Enfoque UTE **11**(1), 1–14 (2019). https://doi.org/10.29019/enfoque.v11n1.586
8. Zaidi, S., Ansari, M.S., Aslam, A., Kanwal, N., Asghar, M., Lee, B.: A survey of modern deep learning based object detection models. Digit. Signal Process. **126**, 103514 (2022). https://doi.org/10.48550/arXiv.2104.11892
9. Xu, R., et al.: ApproxDet: content and contention-aware approximate object detection for mobiles. In: Proceedings of the 18th Conference on Embedded Networked Sensor Systems, pp. 449–462. ACM, Japan (2020). https://doi.org/10.1145/3384419.3431159
10. Zou, Z., Chen, K., Shi, Z., Guo, Y., Ye, J.: Object detection in 20 years: a survey. Proc. IEEE **111**(3), 257–276 (2023). https://doi.org/10.1109/JPROC.2023.3238524
11. Estrada, J., Paheding, S., Yang, X., Niyaz, Q.: Deep-learning-incorporated augmented reality application for engineering lab training. Appl. Sci. **12**(10), 5159 (2022). https://doi.org/10.3390/app12105159
12. Shafer, S.A., Krumm, J., Brumitt, B., Meyers, B., Czerwinski, M., Robbins, D.C.: The New EasyLiving Project. Microsoft Research (1999)
13. Bagchi, S., et al.: Vision paper: grand challenges in resilience: autonomous system resilience through design and runtime measures. IEEE Open J. Comput. Soc. **1**, 155–172 (2020). https://doi.org/10.48550/arXiv.1912.11598

CB-RISE: Improving the RISE Interpretability Method Through Convergence Detection and Blurred Perturbations

Oscar Stanchi[1,4]($\boxtimes$), Franco Ronchetti[1,2], Pedro Dal Bianco[1,3], Gastón Rios[1,3], Santiago Ponte Ahon[1,3], Waldo Hasperué[1,2], and Facundo Quiroga[1,2]

[1] Instituto de Investigación en Informática LIDI - Universidad Nacional de La Plata, La Plata, Argentina
ostanchi@lidi.info.unlp.edu.ar
[2] Comisión de Investigaciones Científicas de la Pcia. de Bs. As. (CIC-PBA), La Plata, Argentina
[3] Universidad Nacional de La Plata (UNLP), La Plata, Argentina
[4] Consejo Nacional de Investigaciones Científicas y Técnicas (CONICET), La Plata, Argentina

Abstract. This paper presents significant advancements in the RISE (Randomized Input Sampling for Explanation) algorithm, a popular black-box interpretability method for image data. RISE's main weakness lies on the large number of model evaluations required to produce the importance heatmap. Furthermore, RISE's strategy of occluding image regions with black patches is not advisable, as it may lead to unexpected predictions. Therefore, we introduce two new versions of the algorithm, **C-RISE** and **CB-RISE**, each incorporating novel features to address the two major challenges of the original implementation. **C-RISE** introduces a convergence detection based on the Welford algorithm which reduces the computational burden of the algorithm by ceasing computations once the importance map stabilizes. **CB-RISE**, additionally, introduces the use of blurred masks as perturbations, equivalent to applying Gaussian noise, as opposed to black patches. This allows for a more nuanced representation of the model's decision-making process. Our experimental results demonstrate the effectiveness of these improvements, successfully enhancing the effectiveness of the generated heatmaps while improving their quality, qualitatively, and showing a speedup of approximately 3.

Keywords: Black Box · Blurred Masks · Computer Vision · Convergence Detection · Deep Learning · Interpretability · RISE

1 Introduction

Deep Learning models are often seen as black boxes due to their complex nature and lack of interpretability [6]. However, the interpretability of these models is gaining importance, with research focusing on understanding the rationale behind their outputs [4,6]. This understanding is crucial for building trust among users and stakeholders [11]. As such, interpretability is becoming a requirement in many tasks, particularly for high-risk decision-making systems [6,12].

Amongst the various black-box interpretability methods, a popular approach for image data is the RISE (Randomized Input Sampling for Explanation) algorithm. RISE, functioning as a post-hoc and local approach tailored for black box models, provides high quality heatmaps indicating the importance of each pixel of a specific input image in a prediction. The importance maps are calculated by repeatedly running the model with masked inputs and weighting the masks according to the model's output. Given an example and a model, the RISE algorithm generates importance maps as depicted in Fig. 1. Our examples use classes based on animals such as an Afghan Hound, an African Hunting Dog, a Flamingo, and a Hog, with the model predicting the correct class with over 96% accuracy in all four cases. The resulting importance maps are normalized in the range between 0 and 1 for better comprehension, where a pixel value of 1 indicates more importance, and viceversa.

An importance map serves as a visualization tool, commonly presented as a heatmap, employed in interpretability methodologies for deep learning models. It illuminates the critical regions within an input image that drive the model's decision-making.

RISE lacks a convergence detection mechanism, which requires users to specify a number of iterations which must be determined beforehand, often leading to a trial-and-error approach. Also, the number of iterations may vary for each image and model due to the dynamic nature of the process, challenging its application to real-word problems. In previous experiment with ImageNet images, this process required thousands of model evaluations [15]. Therefore, the original RISE poses a significant computational burden, rendering the algorithm prohibitively expensive, specially for large models or real time applications.

Furthermore, noise can affect RISE, impacting the accuracy of the generated importance maps. Due to the randomness of the masks and the potential extreme perturbations caused by black patches in the images, the resulting importance maps may not always perfectly reflect the significance of each pixel.

In this article, we propose **CB-RISE**, a modified version of RISE that improves the efficiency and quality of the importance maps. First by utilizing blurred masks we can maintain the spatial distribution of pixels in the masked region. Second, we propose a convergence detection system. Designed to cease computations once the importance map stabilizes, employing the Welford online algorithm for variance calculation. The refinement aims to substantially improve both the implementation for developers and its applicability for end-users, elevating the efficiency of the algorithm in generating importance maps and enhancing the interpretability of black box neural networks.

This article is organized as follows. Section Sect. 1 provides an introduction to the topic and a review of the related work, outlining the motivation and objectives of the research. Section 2 presents our proposed methods, starting with

CB-RISE: Improving the RISE Interpretability Method Through Convergence Detection and Blurred Perturbations

Oscar Stanchi[1,4]([✉])[ID], Franco Ronchetti[1,2][ID], Pedro Dal Bianco[1,3][ID], Gastón Rios[1,3][ID], Santiago Ponte Ahon[1,3][ID], Waldo Hasperué[1,2][ID], and Facundo Quiroga[1,2][ID]

[1] Instituto de Investigación en Informática LIDI - Universidad Nacional de La Plata, La Plata, Argentina
ostanchi@lidi.info.unlp.edu.ar
[2] Comisión de Investigaciones Científicas de la Pcia. de Bs. As. (CIC-PBA), La Plata, Argentina
[3] Universidad Nacional de La Plata (UNLP), La Plata, Argentina
[4] Consejo Nacional de Investigaciones Científicas y Técnicas (CONICET), La Plata, Argentina

Abstract. This paper presents significant advancements in the RISE (Randomized Input Sampling for Explanation) algorithm, a popular black-box interpretability method for image data. RISE's main weakness lies on the large number of model evaluations required to produce the importance heatmap. Furthermore, RISE's strategy of occluding image regions with black patches is not advisable, as it may lead to unexpected predictions. Therefore, we introduce two new versions of the algorithm, **C-RISE** and **CB-RISE**, each incorporating novel features to address the two major challenges of the original implementation. **C-RISE** introduces a convergence detection based on the Welford algorithm which reduces the computational burden of the algorithm by ceasing computations once the importance map stabilizes. **CB-RISE**, additionally, introduces the use of blurred masks as perturbations, equivalent to applying Gaussian noise, as opposed to black patches. This allows for a more nuanced representation of the model's decision-making process. Our experimental results demonstrate the effectiveness of these improvements, successfully enhancing the effectiveness of the generated heatmaps while improving their quality, qualitatively, and showing a speedup of approximately 3.

Keywords: Black Box · Blurred Masks · Computer Vision · Convergence Detection · Deep Learning · Interpretability · RISE

1 Introduction

Deep Learning models are often seen as black boxes due to their complex nature and lack of interpretability [6]. However, the interpretability of these models is gaining importance, with research focusing on understanding the rationale behind their outputs [4,6]. This understanding is crucial for building trust among users and stakeholders [11]. As such, interpretability is becoming a requirement in many tasks, particularly for high-risk decision-making systems [6,12].

Amongst the various black-box interpretability methods, a popular approach for image data is the RISE (Randomized Input Sampling for Explanation) algorithm. RISE, functioning as a post-hoc and local approach tailored for black box models, provides high quality heatmaps indicating the importance of each pixel of a specific input image in a prediction. The importance maps are calculated by repeatedly running the model with masked inputs and weighting the masks according to the model's output. Given an example and a model, the RISE algorithm generates importance maps as depicted in Fig. 1. Our examples use classes based on animals such as an Afghan Hound, an African Hunting Dog, a Flamingo, and a Hog, with the model predicting the correct class with over 96% accuracy in all four cases. The resulting importance maps are normalized in the range between 0 and 1 for better comprehension, where a pixel value of 1 indicates more importance, and viceversa.

An importance map serves as a visualization tool, commonly presented as a heatmap, employed in interpretability methodologies for deep learning models. It illuminates the critical regions within an input image that drive the model's decision-making.

RISE lacks a convergence detection mechanism, which requires users to specify a number of iterations which must be determined beforehand, often leading to a trial-and-error approach. Also, the number of iterations may vary for each image and model due to the dynamic nature of the process, challenging its application to real-word problems. In previous experiment with ImageNet images, this process required thousands of model evaluations [15]. Therefore, the original RISE poses a significant computational burden, rendering the algorithm prohibitively expensive, specially for large models or real time applications.

Furthermore, noise can affect RISE, impacting the accuracy of the generated importance maps. Due to the randomness of the masks and the potential extreme perturbations caused by black patches in the images, the resulting importance maps may not always perfectly reflect the significance of each pixel.

In this article, we propose **CB-RISE**, a modified version of RISE that improves the efficiency and quality of the importance maps. First by utilizing blurred masks we can maintain the spatial distribution of pixels in the masked region. Second, we propose a convergence detection system. Designed to cease computations once the importance map stabilizes, employing the Welford online algorithm for variance calculation. The refinement aims to substantially improve both the implementation for developers and its applicability for end-users, elevating the efficiency of the algorithm in generating importance maps and enhancing the interpretability of black box neural networks.

This article is organized as follows. Section Sect. 1 provides an introduction to the topic and a review of the related work, outlining the motivation and objectives of the research. Section 2 presents our proposed methods, starting with

(a) Afghan Hound

(b) African Hunting Dog

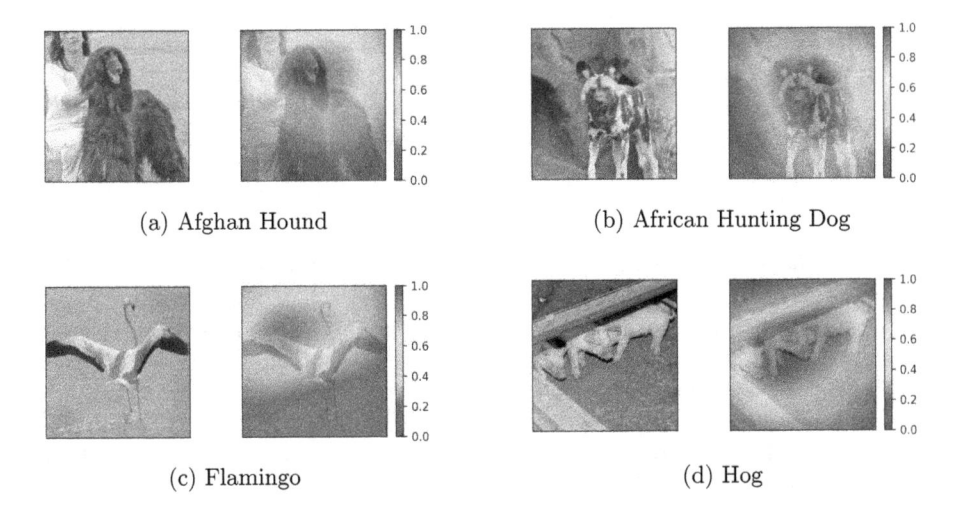

(c) Flamingo

(d) Hog

Fig. 1. Importance Maps generated by RISE using a mask size of 4 × 4. Note that for some cases, such as (b) the importance map clearly reflects the most relevant region, while for (c) the method assigns a large amount of importance to the background.

a background on the original RISE, then we introduce **C-RISE**, a convergence detection mechanism for RISE, followed by **B-RISE**, which integrates blurred masks into the framework. In Sect. 3, we detail the experiments conducted to evaluate our methods to obtain the qualitative and quantitative results. Finally, Sect. 4 concludes the paper, summarizing our findings and suggesting avenues for future research.

1.1 Related Work

Some models, like decision trees or linear/logistic regression, inherently offer interpretability [2,12]. Others require post-hoc analysis for understanding [4,6]. Interpretability methods, local or global, model-specific or agnostic, aid in diagnosing model contributions [5,8]. This understanding is crucial for model improvement and user feedback [1,5].

Recent years have seen the development of methods that train traditionally opaque models in a manner that makes their internal representations more understandable. These advances in this field includes not only intrinsic interpretability, like Concept Whitening [2], where models are trained to obtain an interpretable latent space, but also post-hoc interpretability, where explanations are derived from the model's output after training [4,6]. Both approaches play crucial roles in enhancing our understanding of complex models.

Among the earliest specific techniques in this field were LIME [11] and SHAP [7]. These methods provide straightforward explanations without constraining the capabilities of the original models. However, they overlook the presence of hidden layers, thereby failing to account for the model's precise performance [5,8].

Gradient-based solutions like Saliency Maps, Integrated Gradients, and Grad-CAM offer diverse ways to interpret models. Saliency Maps compute the spatial support of a specific class in an image (image-specific class saliency map) using a single back-propagation pass through a classification CNN, based on calculating the gradient of the class score relative to the input image [14]. Integrated Gradients, on the other hand, require no alterations to the original network and only a few gradient computations. This method adheres to two key axioms: Sensitivity and Implementation Invariance, which attribution methods should satisfy [16]. Grad-CAM, another gradient-based method, employs the gradients of the target class entering the final convolutional layer to generate a rough localization map, emphasizing the image regions crucial for predicting the target class [13].

In addition to these, there exist a variety of perturbation-based approaches that provide alternative avenues for interpreting models. Occlusion is a technique that methodically blocks various parts of the input image to observe the effect on the model's output [17]. Moreover, the RISE algorithm has emerged as a significant development in realm of this type of field [9].

Despite its effectiveness, a notable limitation of RISE is its computational expense, and it lacks the flexibility to use other types of masked-based perturbations, such as blurring [3]. Also, an extension of RISE, known as D-RISE, has been proposed to generate visual explanations for the predictions of object detectors [10]. D-RISE utilizes a proposed similarity metric that considers both localization and categorization aspects of object detection, enabling it to produce saliency maps that highlight the most influential image areas for the prediction. However, it's important to note that D-RISE inherits the same limitations as RISE, including computational expense and lack of perturbation flexibility.

2 Proposed Methods

2.1 Background: Original RISE

RISE operates as a perturbation-based method for computing attribution, utilizing Monte-Carlo integration for averaging model scores over randomly generated masks. In particular, the backbone of the RISE algorithm lies in a specific formulation. For a given input pixel position λ, the output is represented as:

$$S_{I,f}(\lambda) \stackrel{MC}{\approx} \frac{1}{\mathbb{E}[M] \cdot N} \sum_{i=1}^{N} f(I \odot M_i) \cdot M_i(\lambda) \tag{1}$$

Here, $S_{I,f}$ denotes the importance map, which elucidates the decision made by model f on the input image I. This map is normalized to the expectation of M, a random binary bilinear upsampled mask set of N elements.

In the context of the RISE algorithm, masks sizes play a crucial role in determining the level of detail captured in the importance maps. As illustrated in Fig. 2, Larger masks lead to more concentrated importance patches, which can significantly impact the efficacy of the method, particularly when the image

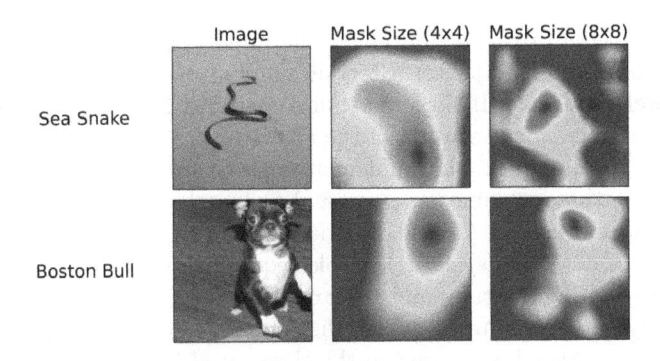

Fig. 2. Comparison of importance maps generated by the RISE algorithm with different mask sizes after 4096 iterations.

contains small objects or fine details. Conversely, smaller masks may be more suitable when the image features larger objects or patterns.

Furthermore, the number of masks used directly influences the reliability of the results. Using a larger number of masks leads to a more reliable convergence of the algorithm. However, it's important to note that increasing the number of masks implies a linear increase in the cost of the computation process. Therefore, the choice of mask size and number should be guided by the specific characteristics of the problem being addressed, balancing the need for detail and computational efficiency.

2.2 C-RISE: Convergence Detection Mechanism for RISE

One notable challenge is the requirement for a significant number of masks to compute the importance maps efficiently. To address this challenge, we propose the implementation of a convergence detection mechanism aimed at reducing the number of masks needed to conclude. In this section, we explore the potential application of the Welford algorithm as a solution for achieving convergence of importance maps in a more efficient manner.

To accomplish our goal, we have incorporated three parameters within the attribute method to our implementation:

- `patience`: This is the number of iterations with no changes after which the execution will be stopped.
- `d_epsilon`: This is the minimum change between the variances of two successive importance maps (separated by `patience` iterations) that is required for a certain pixel to be counted as having converged.
- `threshold`: This is the proportion of pixels in the importance map that are allowed to change. The computation of the importance map stops when the proportion of converged pixels (`1 - threshold`) is reached.

In the pursuit of algorithm convergence, our focus lies in assessing the stability of the generated importance map. To achieve this objective, we monitor the variation in the pixel distribution of the importance map (calculated using Welford's online algorithm) after a certain number of iterations, determined by the parameter `patience`. If the importance map has reached a state of convergence, we terminate the execution of the algorithm.

Within this framework, the implementation of the Welford algorithm only requires storing three variables: `n`, `mean` and `std`, and it calculates the mean and variance simultaneously, as desired. Furthermore, in the context of implementing the convergence detection system, the sole requirement is to retain the variance for both the resulting `heatmap` and the `previous_heatmap`. This approach optimizes memory usage and improves the efficiency of our algorithm without adding computational overhead compared to the time saved by its early termination[1].

2.3 B-RISE: Blurred Mask Integration for RISE

In our previous work [15], we identified a potential area of study concerning the impact of black patches from masks, which are used to obscure regions of an image, on the output score of the model. We propose instead of using black patches, to blur that region of the image, or average the colors of the masked section.

In this process, we consider an input image I and a binary continuous mask $M_i \in M$. The mask, which ranges from values of zero to one, is upsampled using bilinear interpolation. The masked blurred input image I_M is then achieved by applying the following formula:

$$I_M = M_i \odot I + (1 - M_i) \odot I_B \qquad (2)$$

where I_B denotes the blurred input image obtained through the application of a Gaussian filter with standard deviation σ. This process involves a single convolution of the input image, thereby not compromising the performance.

This formula essentially applies the mask to the input image and inversely to the blurred input image, combining the two results to produce the final output image. The mask serves to control the degree of blurring in the output image: regions of the image where the mask is closer to one will resemble the input image, while regions where the mask is closer to zero will be more blurred. This approach allows for a smooth transition between the original and blurred regions of the image, providing a more natural-looking result. Figure 3 illustrates the process of applying this formula visually.

The selection of the standard deviation for the Gaussian filter is highly dependent on the application, and as such, there is no definitive rule. However, typically, it is advisable for the user to choose a σ value for the Gaussian filter that results in a significant amount of high-frequency components in the image being filtered out. Figure 4 illustrates this relationship, showing how blur intensity changes with varying σ values, with examples for values of 2.5, 10, 50 and ∞

---

[1] The complete implementation, along with the entire codebase developed for this project, is available in the project's repository: https://github.com/indirivacua/cbrise/tree/main.

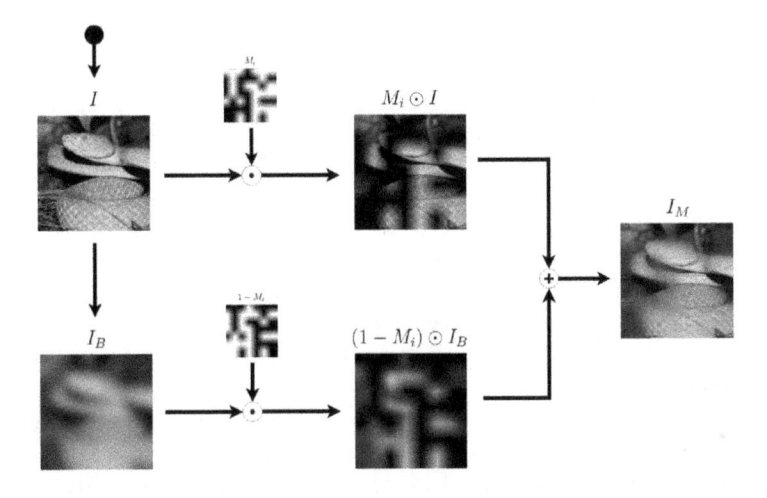

Fig. 3. B-RISE's approach to masking inputs.

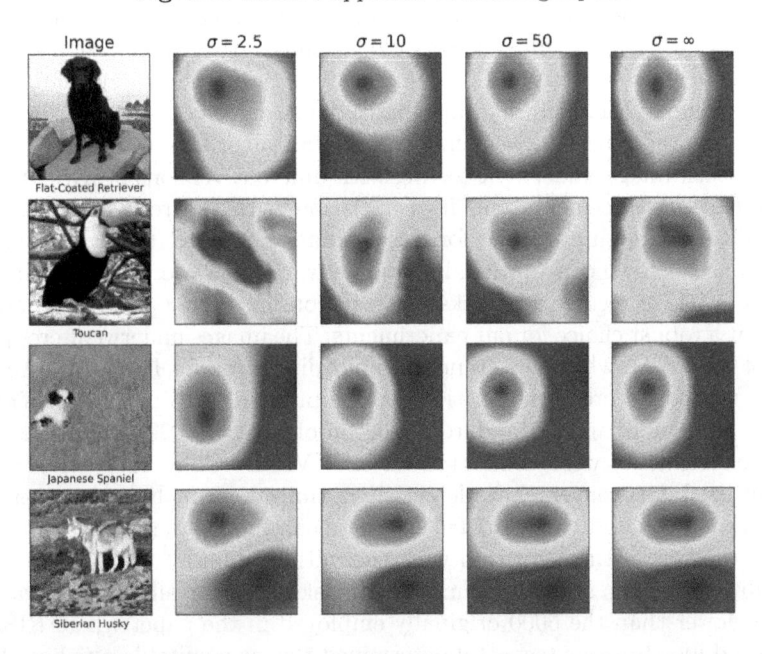

Fig. 4. Illustration of the relationship between blur intensity and varying σ values. Examples are provided for σ values of 2.5, 10, 50, and ∞ (equivalent to masking with a costant value), after 4096 iterations with a mask size of 4×4.

(representing an infinitely large kernel radius where the blur effectively averages the colors of the masked section across the entire image).

3 Experiments and Results

In this section, we describe the experiments we performed with the proposed methods, comparing them to the original RISE. We present both **qualitative** and **quantitive** results as is typical for interpretability methods.

3.1 Experiments

The evaluation focused on assessing the effectiveness of the RISE algorithm variations in enhancing interpretability and feature attribution in the context of image classification tasks, while also considering their improvements in computational efficiency.

In our preceding study [15], we observed that the computational load scales linearly with the increase in the number of masks, as each additional mask requires extra computational workload. Interestingly, computational time remains unaffected by mask size, serving solely to modulate the level of detail captured within the importance maps. Consequently, our analytical focus zeroes in on mask quantity, enabling a meticulous comparative examination across different algorithmic versions and its configurations, alongside an evaluation of the coherence among the resulting importance maps.

We evaluated our improvements in the RISE algorithm using the ImageNet dataset. Each image underwent testing with different versions of RISE to assess performance enhancements and its correlation with the resulting importance maps. The architecture utilized for this evaluation is VGG16, a widely adopted and very well-known convolutional neural network known for its depth and effectiveness in image recognition tasks. VGG16 consists of 138,357,544 parameters, making it a robust choice for our experiments. The images underwent cropping to a size of 224 × 224, which is a standard input dimension for ImageNet. Additionally, we applied z-score normalization to the input images using a mean of `mean` = $[0.485, 0.456, 0.406]$ and a standard deviation of `std` = $[0.229, 0.224, 0.225]$.

All experiments were conducted on an NVIDIA GTX 1060 GPU. Parameters utilized for convergence detection optimization included `patience=64`, `d_epsilon=1e-3`, and `threshold=0.3`. These parameters were applied uniformly to both versions of the algorithm with and without blur.

Additionally, the standard number of masks used for all versions was 4096, which is lower than the 6000 originally employed in the paper where RISE was introduced [9]. Previous tests [15] determined this as a suitable number of iterations to obtain a reliable heatmap, surpassing the convergence threshold of the algorithm consistently across various problem types and mask sizes.

The SSIM (Structural Similarity Index) metric was used to assess the similarity between the modified methods' importance maps and the importance map from the original RISE implementation. SSIM measures the similarity between two images, considering luminance, contrast, and structure. Higher SSIM values indicate greater similarity between images. In our case, SSIM was used to evaluate the similarity between importance maps generated by different methods, with higher values indicating closer resemblance.

3.2 Results

Qualitative Results. Based on our results from Fig. 5a, the three versions of the RISE algorithm executed with a mask size of 4 × 4 and 4096 masks exhibit variations in the number of iterations required for importance map generation. The original version exhaustively employed all masks in the iterative process, since it does not have a convergence criterion. On the other hand, the other two versions, equipped with the convergence detection mechanism, completed significantly fewer iterations (consistently fewer than 1600). This reduction in the number of iterations highlights the efficacy of the new improvements, particularly the incorporation of a convergence detection approach, in reducing the computational burden of the algorithm. Despite this discrepancy in computational efficiency, a notable consistency emerges in the generated importance maps across all versions. Moreover, these robust results of the algorithm in capturing and highlighting salient features within the images are achieved with fewer iterations due to the prior convergence of the algorithm facilitated by the implementation of the mechanism. Thus, while the efficiency of the algorithm may vary depending on the version utilized, the quality and coherence of the generated importance maps remain commendably consistent.

In Fig. 5b, it can be observed that when using 4096 masks of size 8 × 8, the results are largely consistent with those obtained from the 4 × 4 mask size. However, due to the increased number of possibilities (i.e., $2^{8\times 8}$), the resulting importance map exhibits slightly more noise. This is manifested as additional artifacts in the importance map. Despite this, all versions of the RISE algorithm, including those equipped with convergence detection, continue to generate importance maps that capture and highlight salient features within the images. The incorporation of the convergence detection strategy remains effective in reducing the computational burden of the algorithm, even with the increased mask size and count, since in this case fewer than 1600 iterations were needed to reach the stabilization of the importance map with and without blurring. Therefore, while the importance map may be slightly noisier due to the increased number of possibilities, the quality and coherence of the generated importance maps remain commendably consistent across all versions of the RISE algorithm.

When comparing results from different versions of the algorithm in Fig. 5, we notice that the generated importance maps exhibit notable consistency. As the method uses a random generator for masks, the convergence for an example occur stochastically and the importance map generated may not always require the exact same number of iterations to stabilize.

Particularly, for both case of mask size, running **B-RISE** with the same number of iterations as the original RISE provides insight that the differences in the resulting importance maps are primarily attributed to the application of blur rather than the fewer iterations. This is evident because both importance maps generated with **B-RISE** and **CB-RISE** show a notable similarity in the regions considered for the model's decision-making process.

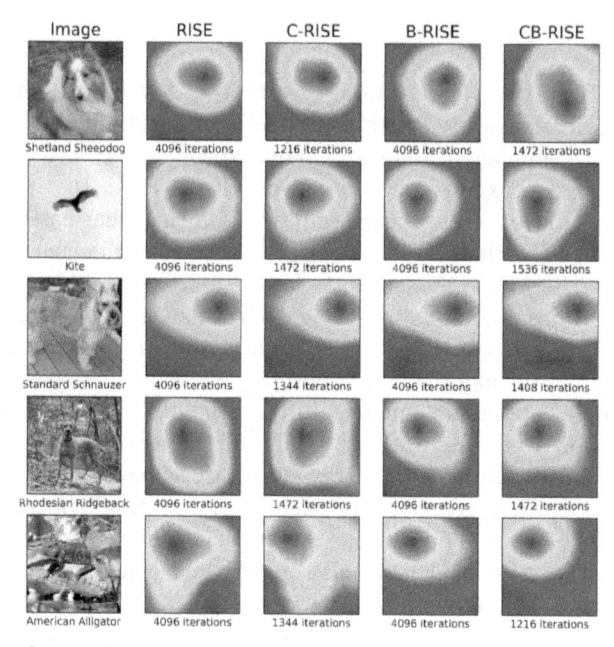

(a) Variations of the RISE algorithm using a standard of 4096 masks and a size of 4×4.

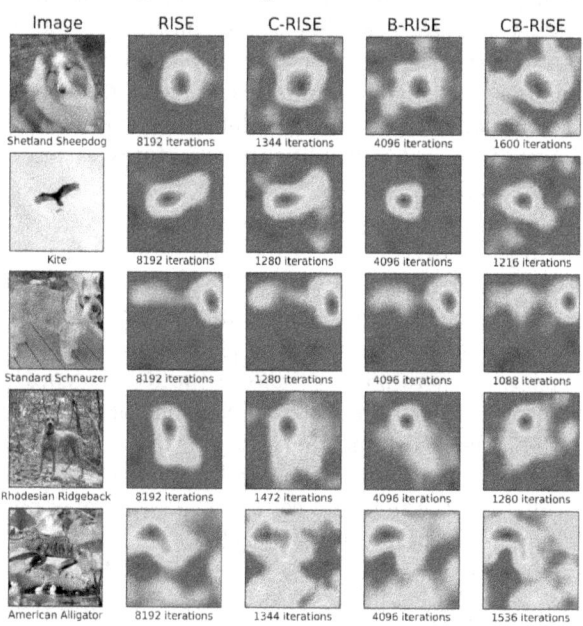

(b) Variations of the RISE algorithm using a standard of 4096 masks and a size of 8×8.

Fig. 5. Variations of the RISE algorithm presented in five columns per image. The first column displays the original image, followed by different RISE versions. The second column features the original RISE, the third integrates convergence detection (**C-RISE**), the fourth applies a blur filter with $\sigma = 10$ (**B-RISE**), and the fifth combines convergence detection with the same blur filter (**CB-RISE**).

Quantitative Results. We evaluate the efficiency of the proposed methods by measuring their speedup with respect to the original RISE method in terms of the number of iterations required. Additionally, to ensure that **C-RISE** does not introduce artifacts or significant differences in the results, we compare the resulting importance maps with respect to the original RISE.

Table 1 summarizes the average and standard deviation for the number of iterations and SSIM measure for the two methods: **C-RISE** and **CB-RISE**, with their respective hyperparameters. These statistics were obtained from evaluating 30 images.

The average number of iterations with a mask size of 4×4 for the **C-RISE** method was found to be 1525.33, while for the **CB-RISE** method, it was 1495.47. For a mask size of 8×8, the average number of iterations for both the **C-RISE** and **CB-RISE** methods was 1388.80. These results provide insights into the computational efficiency of each method and the impact of blur on the convergence behavior.

These findings resulted in a speedup of 2.69 for **C-RISE** with a mask size of 4×4, and 2.74 for 8×8. Furthermore, **CB-RISE** exhibited an even greater speedup, reaching 2.95, significantly enhancing the efficiency of the method.

Intuitively, one might expect that larger mask sizes would result in more detail, leading to more noise, which decreases with more iterations. However, the average number of iterations for 8×8 masks was lower than for 4×4 masks for both methods. This is compensated by a higher standard deviation compared to that of 8×8 masks.

Additionally, given that the mean values of SSIM are closer to 1, the resulting importance maps are similar to those generated by the original RISE method, indicating that important features are preserved across methods. This ensures that the modifications introduced do not compromise the effectiveness of the generated importance maps.

Table 1. Average and standard deviation number of iterations for **C-RISE** and **CB-RISE** methods, along with SSIM values, compared to the original RISE importance map. The experiments were conducted using a VGG16 architecture and a dataset comprising 30 images from ImageNet.

Measure	Aggregation	RISE (4×4)	RISE (8×8)	C-RISE (4×4)	C-RISE (8×8)	CB-RISE (4×4)	CB-RISE (8×8)
Iterations	Avg	4096	4096	1525.34	1388.80	1495.46	1388.80
	Std	0	0	116.54	137.73	168.48	174.79
SSIM	Avg	1	1	0.92	0.81	0.88	0.75
	Std	0	0	0.06	0.11	0.06	0.10

4 Conclusions and Future Work

4.1 Conclusions

We have made significant strides in enhancing the efficiency and effectiveness of the RISE algorithm. By introducing blurred masks with **B-RISE** and a convergence detection system in **C-RISE**, we have addressed two major challenges associated with the original implementation of the algorithm.

The use of blurred masks, equivalent to applying Gaussian noise to the masked portions of the image, as opposed to replacing them with black patches, has allowed for a more nuanced representation of the model's decision-making process. Our findings suggest that occluding image regions with black patches is not advisable. It is possible that black patches may deactivate chains of neural activations, leading to the filtering of classes and potentially influencing the expected class prediction, resulting in unexpected or odd predictions.

This approach has proven to be effective in our experiments in preserving the spatial distribution of pixels in the occluded region, decreasing the volume of information without creating odd edges and textures in the image, thereby providing a more accurate importance map with fewer artifacts. This aligns with the assumptions of the original RISE implementation regarding the selection of occlusion masks, which assumes a strong spatial structure in the feature space, implying that closely situated features are more likely to be correlated, thus justifying the use of blurred masks.

Additionally, using blurred masks does not introduce an overhead in compute because it requires only one convolution of the input image.

Furthermore, the incorporation of a convergence detection mechanism, based on the Welford algorithm has addressed need for manually determining the number of iterations in the RISE algorithm needed to converge. It is worth emphasizing that each image may require a different number of iterations due to the dynamic nature caused by the randomness of the generator of masks, the model's training, and the image itself.

This addition has not only reduced the computational burden of the algorithm but also improved its efficiency by ceasing computations once the importance map converges. Despite the variations in computational efficiency across different versions of the RISE algorithm, the quality and coherence of the generated importance maps have remained commendably consistent.

Our experimental results have demonstrated the effectiveness of these improvements. The average number of iterations for the **C-RISE** method has significantly decreased compared to previous implementation that used a fixed number for the same model without regard for the specific image, showcasing a notable improvement in computational efficiency. For the **CB-RISE** method, we show an even larger reduction in the average number of iterations. This reduction suggests a more streamlined convergence behavior, indicating the positive impact of incorporating blur into the occlusion masks. This consistency has been maintained even when the number of masks and their size were increased, further attesting to the robustness of our proposed modifications. Additionally, the similarity between the resulting importance maps and those generated by the original RISE method, as measured by SSIM, demonstrates that important features are preserved across methods. These findings affirm that the modifications introduced successfully enhance the effectiveness of the generated importance maps without compromising their quality.

4.2 Future Work

One potential direction for future investigation involves exploring alternative masking strategies, such as masking the output (activation maps) of a convolutional layer, particularly the final layer. By focusing solely on the classification part of the network, this approach could potentially reduce computational complexity, since computations for previous layers of the network do not need to be performed, only those chosen by the user.

Another interesting path for future research is to leverage feature maps to generate the mask distributions. We believe that by incorporating feature maps directly into the mask generation process, we may achieve more refined and context-aware interpretations.

Finally, exploring metaheuristic algorithms such as evolutionary approaches for mask generation could be rewarding for reducing the computational cost of the algorithm. Starting with a set of randomly generated masks, the algorithm could iteratively select and refine the most promising candidates, thus improving the quality of the importance map generated and the efficiency of the method.

References

1. Broniatowski, D.A., et al.: Psychological foundations of explainability and interpretability in artificial intelligence. Technical report, NIST (2021)
2. Chen, Z., Bei, Y., Rudin, C.: Concept whitening for interpretable image recognition. Nat. Mach. Intell. **2**(12), 772–782 (2020)
3. Dabkowski, P., Gal, Y.: Real time image saliency for black box classifiers. In: Advances in Neural Information Processing Systems, vol. 30 (2017)
4. Doshi-Velez, F., Kim, B.: Towards a rigorous science of interpretable machine learning. arXiv preprint arXiv:1702.08608 (2017)
5. Escalante, H.J., et al.: Explainable and Interpretable Models in Computer Vision and Machine Learning. Springer, Heidelberg (2018). https://doi.org/10.1007/978-3-319-98131-4
6. Lipton, Z.C.: The mythos of model interpretability: in machine learning, the concept of interpretability is both important and slippery. Queue **16**(3), 31–57 (2018)
7. Lundberg, S.M., Lee, S.I.: A unified approach to interpreting model predictions. In: Advances in Neural Information Processing Systems, vol. 30 (2017)
8. Molnar, C., Casalicchio, G., Bischl, B.: Interpretable machine learning – a brief history, state-of-the-art and challenges. In: Koprinska, I., et al. (eds.) ECML PKDD 2020. CCIS, vol. 1323, pp. 417–431. Springer, Cham (2020). https://doi.org/10.1007/978-3-030-65965-3_28
9. Petsiuk, V., Das, A., Saenko, K.: RISE: randomized input sampling for explanation of black-box models. In: Proceedings of the British Machine Vision Conference (BMVC) (2018)

10. Petsiuk, V., et al.: Black-box explanation of object detectors via saliency maps. In: Proceedings of the IEEE/CVF Conference on Computer Vision and Pattern Recognition, pp. 11443–11452 (2021)
11. Ribeiro, M.T., Singh, S., Guestrin, C.: "Why should i trust you?" Explaining the predictions of any classifier. In: Proceedings of the 22nd ACM SIGKDD International Conference on Knowledge Discovery and Data Mining, pp. 1135–1144 (2016)
12. Rudin, C.: Stop explaining black box machine learning models for high stakes decisions and use interpretable models instead. Nat. Mach. Intell. **1**(5), 206–215 (2019)
13. Selvaraju, R.R., Cogswell, M., Das, A., Vedantam, R., Parikh, D., Batra, D.: Grad-CAM: visual explanations from deep networks via gradient-based localization. In: Proceedings of the IEEE International Conference on Computer Vision, pp. 618–626 (2017)
14. Simonyan, K., Vedaldi, A., Zisserman, A.: Deep inside convolutional networks: visualising image classification models and saliency maps. arXiv preprint arXiv:1312.6034 (2013)
15. Stanchi, O., Ronchetti, F., Quiroga, F.: The implementation of the RISE algorithm for the captum framework. In: Naiouf, M., Rucci, E., Chichizola, F., De Giusti, L. (eds.) JCC-BD&ET 2023. CCIS, vol. 1828, pp. 91–104. Springer, Cham (2023). https://doi.org/10.1007/978-3-031-40942-4_7
16. Sundararajan, M., Taly, A., Yan, Q.: Axiomatic attribution for deep networks. In: International Conference on Machine Learning, pp. 3319–3328. PMLR (2017)
17. Zeiler, M.D., Fergus, R.: Visualizing and understanding convolutional networks. In: Fleet, D., Pajdla, T., Schiele, B., Tuytelaars, T. (eds.) ECCV 2014, Part I. LNCS, vol. 8689, pp. 818–833. Springer, Cham (2014). https://doi.org/10.1007/978-3-319-10590-1_53

Wavelength Calibration of Historical Spectrographic Plates with Dynamic Time Warping

Santiago Andres Ponte Ahón[1,2]([✉])[iD], Juan Martín Seery[1] [iD], Facundo Quiroga[1,3] [iD], Franco Ronchetti[1,3] [iD], Oscar Stanchi[1] [iD], Pedro Dal Bianco[1,2] [iD], Waldo Hasperué[1] [iD], Yael Aidelman[4,5] [iD], and Roberto Gamen[5] [iD]

[1] III-LIDI, Facultad de Informática, UNLP, 1900 La Plata, Buenos Aires, Argentina
sponte@lidi.info.unlp.edu.ar
[2] UNLP, La Plata, Argentina
[3] Comision de Investigaciones Científicas de la Pcia. De Bs. As. (CIC-PBA), La Plata, Argentina
[4] Facultad de Ciencias Astronómicas y Geofísicas, UNLP, La Plata, Argentina
[5] Instituto de Astrofísica de La Plata, CONICET-UNLP, La Plata, Argentina

Abstract. The *Facultad de Ciencias Astronómicas y Geofísicas* of the *Universidad Nacional de La Plata* counts with 15,000 spectroscopic records on glass plates with valuable and unique astronomical data.

Currently, processing these plates requires a complex manual process that involves several stages, requiring several hours to process a single plate. In particular, the wavelength calibration requires the determination of the wavelength range in which the data were observed. This is achieved by matching the spectrum of the comparison lamp on the plate to the reference spectrum. Since many times neither the metadata of the lamps nor the physical lamps are available, automating the tasks requires a semi-blind approach that uses simulated data as a reference. However, the simulated data differs significantly from the physical lamps, given that many peaks determined by theoretical calculations are rarely observed in practice, and conversely the physical lamps and spectrograph carry imperfections that cause unexpected peaks.

In this work, we propose an wavelength calibration pipeline that enables automated matching of the wavelength of the comparison lamps via *Dynamic Time Warping* (DTW) between the samples and simulated data. Our best model achieves a 93% average Intersection-over-Union (IoU) over a set of 32 manually calibrated plates.

Keywords: Dynamic Time Warping · Spectroscopic records · Spectroscopic plates · Optimization · Wavelength calibration

1 Introduction

Glass spectroscopic plates became popular between 1920 and 1980 to preserve information from stellar observations. The observation methodology consists of capturing stellar light rays, dispersing them through a prism recording the resulting electromagnetic spectrum on a glass plate using a photosensitive reagent.

Currently this methodology has been replaced by modern and efficient alternatives. However, during the time it was booming, a lot of data was generated that still persists today. For example institutions such as Harvard [1] and the Heidelbergel observatory [5] still keep data of these characteristics. Unfortunately, it is particularly difficult to work with the information from these plates once taken. Their treatment is to a certain extent artisanal, requiring highly trained personnel and many man-hours to process the information from a single plate. Furthermore, the plates are highly volatile over time, mainly due to their fragility and storage quality.

The availability of historical data in digital format allows for the comparison of observations from different periods, enabling the detection of differences and variations in astronomical sources over extended periods of time, covering approximately 100 years of observations. It also allows for the analysis of data from several decades ago with modern tools, enabling the contrast of these data with current theoretical models. This implies that new discoveries are latent in the spectroscopic plates, which makes them a highly relevant historical heritage. In 2000 the International Astronomical Union issued resolution *No. B3 'Safeguarding the information in photographic plates'* [14], where they indicate the need to take measures to ensure the conservation of historical data of these characteristics, in order to prevent them from being lost for future generations of astronomers. In 2019, the *Facultad de Ciencias Astronómicas y Geofísicas* (FCAG) of the *Universidad Nacional de La Plata* (UNLP) created the Recuperation of Historic Observational Work (ReTrOH, for its initials in Spanish) project [10] with the goal of preserving and digitizing more than 15,000 of these spectroscopic records, recorded in the southern hemisphere between 1920 and 1980. To date, ReTrOH has digitised in excess of 250 spectroscopic plates, including, for example, spectra from the night of the discovery of Nova Puppis on 8 November 1942 [2].

Under this context, it becomes evident how important it is to find a faster method than the current one to process the information from this type of plates. Processing a single plate can take hours, if not days, of manual processing time with highly specialized personnel and there are hundreds of thousands of these plates in existence. In this work, automatic alternatives are sought to mitigate this situation and make the processing of large collections of old spectroscopic plates a little more viable.

Figure 1 graphically illustrates the various stages involved in the typical processing of a spectroscopic plate. The stage that this work focuses on is highlighted in red.

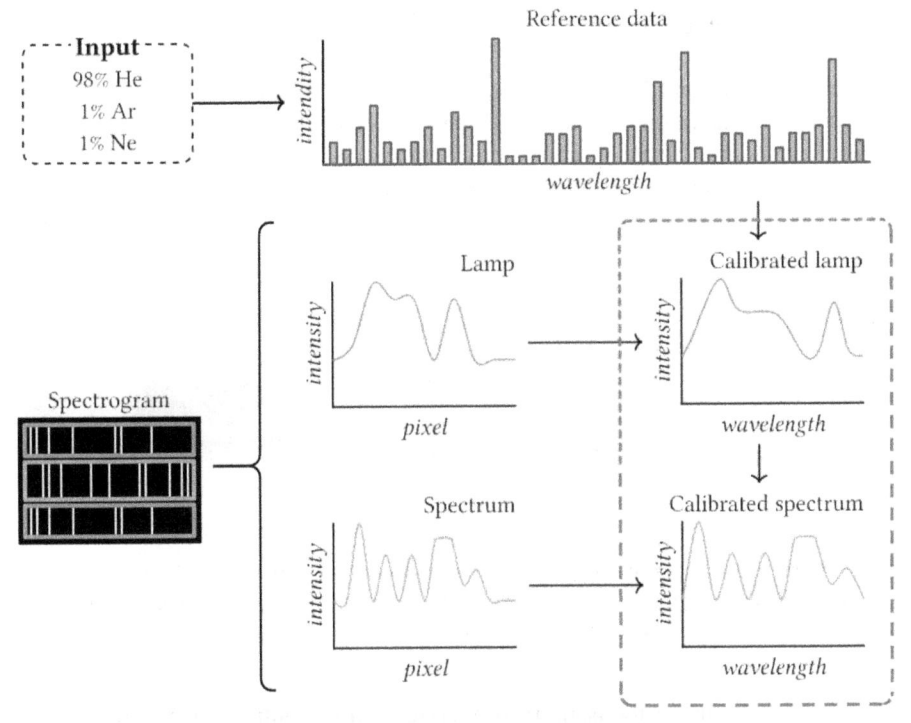

Fig. 1. Diagram of the various stages involved in processing a spectrum scan of a spectrographic plate. It includes the segmentation of each spectrum component and the extraction of the functions that describe each one. It also covers obtaining the components of the lamp used by the user and acquiring the reference data necessary for subsequent calibration of the comparison lamp function. Following this, the wavelength calibration of the main spectrum function is performed. Arrows indicate the resources required for each step. The stages highlighted in red are the focus of this work. (Color figure online)

1.1 Related Works

The majority of institutions have digitised their astrometric or photometric plate collections. However, only two works have been identified that relate to the digitisation of spectroscopic plates.

The goal of PyPlate software [13] is to assist in various stages of the digitization of *photographic* plates. However, it has no support for spectroscopic plate processing, which differ significantly in their analysis.

A notable example is the study carried out by [9], where the treatment of a set of photographic plates from various sources is carried out using PyPlate. This describes in detail how these plates can be treated and processed in order to improve their processing errors. However, the nature of their information is different so it is not feasible to simply apply the procedures that are valid on photographic plates to spectroscopic plates.

In previous works we have already explored how to properly digitize this type of plates and how to automatically identify them in their digital scans [7, 11].

2 Problem Description

Between the 1920s and 1980s, spectroscopic plates are in small glass rectangles with light spectrums etched on their surface. Typically, each plate comes bundled with some metadata annotations from when the observation was made. Figure 2 illustrates the appearance of one of these plates and the envelope in which it is stored.

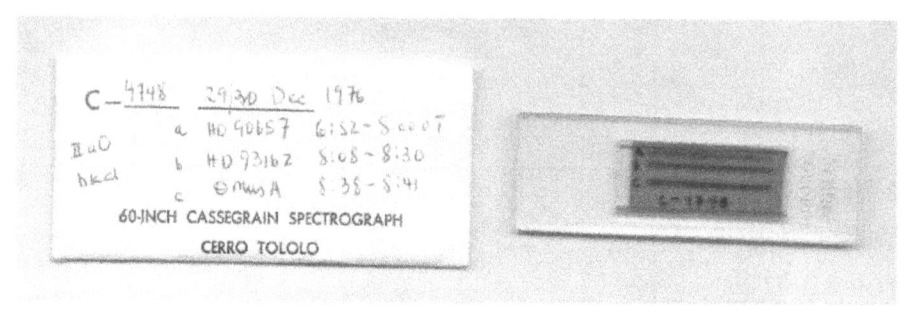

Fig. 2. Right, glass spectroscopic plate with three spectrums on its surface. Left, plate storage envelope, which has various metadata about each spectrum on the plate noted. The spectrums identify them from each other by the references a, b and c respectively.

The initial step in the digitisation process is the high-resolution scanning of the plate. Once the plate has been scanned, the images of each of the spectra found on it must be isolated. For this purpose, we utilise the PlateUNLP software, which we have developed for this specific purpose. This software enables the scanned image of the plate to be cropped in order to isolate each observation, which is then saved in a FITS file along with its corresponding metadata (header) [7, 11]. Each of these FITS files is therefore an image of a portion of the plate containing the spectrum corresponding to the observed object. Each of these "science spectra" is always accompanied by two comparison lamp spectra. Figure 3 shows an example of a FITS file image, each spectrum is indicated with a box. Comparison lamps are strictly necessary to carry out the wavelength calibration of a spectrum;

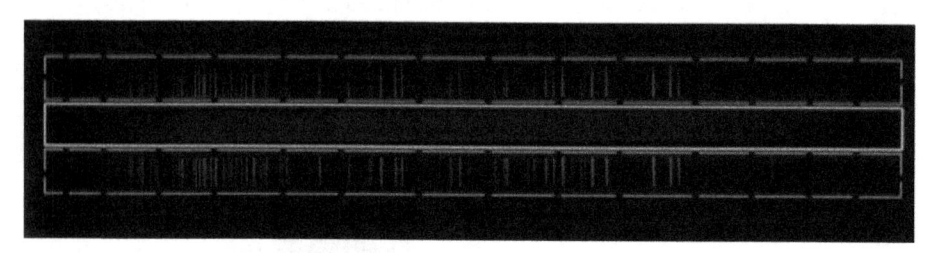

Fig. 3. In the centre, the green box shows the science spectrum. The spectra in the red boxes correspond to the spectra of the comparison lamps. (Color figure online)

In order to obtain the descriptive function of intensities for each spectrum (lamps and science), it is necessary to determine the value of X, which corresponds to the position on a line passing through the centre of the spectrum in the direction of dispersion. The value of I is then calculated as the average of the pixel values in the direction perpendicular to the X-axis. Consequently, three functions are generated (two identical ones describing the spectra of the comparison lamp and one for the spectrum of science), each containing the information pertaining to the average intensity, I, of each pixel. In the field of astronomy, this process is referred to as the extraction of the spectrum and is typically carried out using IRAF tasks [6]. Figure 4 shows a extracted comparison lamp spectrum. The final step is to identify a transformation that will align the x-axis with the wavelength. This process is referred to as wavelength calibration and is discussed in the subsequent section.

2.1 Wavelength Calibration

Once we have the descriptive function of the science spectrum, it is necessary to calibrate them in wavelength. This is where the comparison lamps come into play. A comparison lamp comprises a known mixture of gases. The user is typically aware of the chemical elements (He, Ar, Ne, Fe, etc.) present in the lamps on a plate. The wavelengths corresponding to the intensity peaks (emission lines) observed in its spectrum are also known. Therefore, the lamp spectrum enables the construction of a function $f(x) = \lambda$, where x represents the position (in pixels) in the dispersal direction and λ is the corresponding wavelength. The lamp elements are frequently associated with the spectrograph employed at the time the plates were recorded. The majority of them are iron-based lamps.

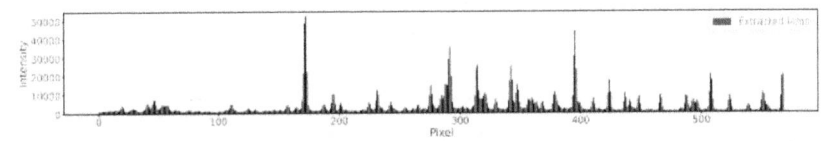

Fig. 4. The extracted comparison lamp spectrum was obtained from a digitised spectroscopic plate. The X axis represents the pixels corresponding to the dispersal direction, while the Y axis corresponds to the average intensity of the pixels perpendicular to the X axis.

The calibration process is done manually, the possible peaks of each chemical element are counted in the hundreds and possibilities have to be explored across lots of possible wavelengths. Also, we must consider the noise that may exist in the analyzed spectra. This process can take a long time and depends mainly on the perseverance, knowledge and ability to identify patterns of the astronomer who is performing the calibration. Figure 5 shows an example of a well-calibrated comparison lamp spectrum.

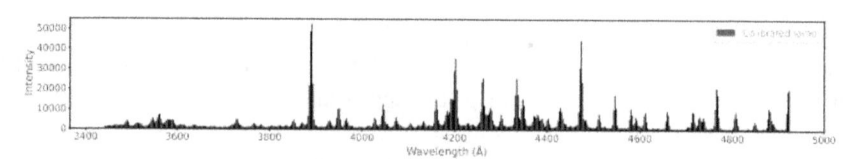

Fig. 5. Comparison lamp with intensities calibrated in wavelength (Å).

With the previously obtained transformation, the problem is practically solved. The transformation is applied to the science spectrum and the result is the wavelength-calibrated spectrum. Figure 6b shows the calibrated spectrum that corresponds to the lamp in the Fig. 5.

Ideally, the processed data is stored/published in some trusted data repository, ensuring their conservation so they can be used in future research.

3 Automatic Wavelength Calibration

In order to improve the processing and calibration times, we carried out an exhaustive analysis of experts manual processing methods. As a result we designed an automatic approach to wavelength calibration. Throughout this section, we describe the approach in detail.

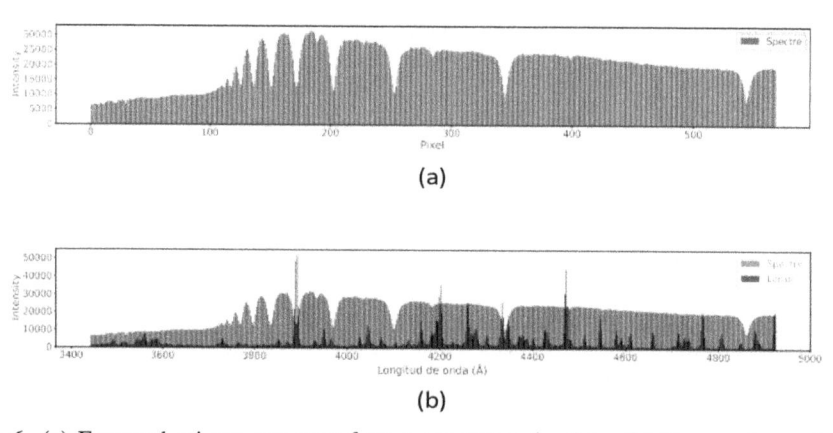

Fig. 6. (a) Extracted science spectrum from a spectroscopic plate. (b) The previous science spectrum calibrated in wavelength (Å).

3.1 Obtaining Reference Data

As explained in the previous section, the chemical elements (He, Ar, etc.) that correspond to the lamps on a plate are generally known to the user. The lamp chemical elements usually depend on the tool with which the plates were recorded at the time. Therefore, the chemical element can be input by the expert user of the software.

Once the chemical elements of interest have been identified, it is necessary to find the spectrum that corresponds to a lamp assembled with these chemical elements. This can be obtained by consulting specialized databases, such as the Atomic Spectra Database (ASD) made by the National Institute of Standards and Technology (NIST) [8]. ASD exposes an interface for Laser-Induced Breakdown Spectroscopy (LIBS), which can be provided information on chemical elements, percentages and range to be considered, the interface can compute the hypothetical spectrum that a lamp would have in the indicated range [4]. Additionally, ASD can also provide only the peaks of the spectra corresponding to each chemical element – with sufficient knowledge of the domain one can combine and process their information to deduce an approximation of what the required spectrum should be like.

In this work, two data sources were mainly considered:

1. NIST: to obtain only the position of the intensity peaks of those chemical elements of interest
2. LIBS: to obtain the same information of the NIST peaks but with the addition that it tries to deduce the intermediate values between the different peaks through a series of specialized smoothing functions. The amount of data deducted depends on the resolution that the user specifies.

3.2 Wavelength Calibration

Once reference data has been obtained, the next steps consists of the wavelength calibration of the observed data with respect to the reference data. To ensure accurate outputs, we must consider a couple of characteristics about the functions of the comparison lamps:

1. The lamps observed can have a non-uniform sampling in the wavelength axis, due to the construction characteristics of the spectrograph. Therefore, they may be slightly irregular with respect to how many angstroms correspond to each pixel.
2. Although the intensities on the Y axis usually maintain their own internal ratio, the maximums and minimums of each individual function can vary greatly from other functions and the reference data. Naturally, this makes it difficult to directly compare the same direct comparison between 2 functions.

To deal with the observed irregularities along the X axis, the well-known time series alignment algorithm "Dynamic Time Wrapping" (DTW) was considered appropriate. This allows for point-to-point correspondences of each value of one function with respect to another. The most interesting feature of it is that it allows deformations to be carried out along the X axis in order to achieve optimal alignment of the values. The degree of deformations allowed is also configurable via a weighting scheme.

To deal with the different ranges in which the Y axis moves, a normalization step was added prior to the application of the DTW algorithm. After some preliminary tests, it was determined that this step was necessary to obtain coherent calibrations between the observed data and the reference data.

For DTW calculations we use the dtw-python library [3, 12].

3.3 Reference Data Segmentation

As mentioned before, the observed function corresponds to a small segment of the reference function (see Fig. 7). Given this situation, it is inappropriate to use DTW to compare the entire spectrum to be analyzed as a reference, since in this case the algorithm would try to fit the observed function into the entire reference. The entire reference lamp usually covers wavelengths from 0 to 20,000 Å and the type of empirical lamps that we want to calibrate tend to occupy a range approximately 3,000 Å wide.

Therefore, a windowing approach was used to solve this problem. Given a range of 20,000 Å, a window is separated every 25 Å, each one occupying a range of 2,000–4,000 Å. The DTW algorithm is applied to each of the windows and the resulting alignment is stored from each execution. Figure 8 shows an illustrative graph of how the segmentation of windows is carried out on the reference data.

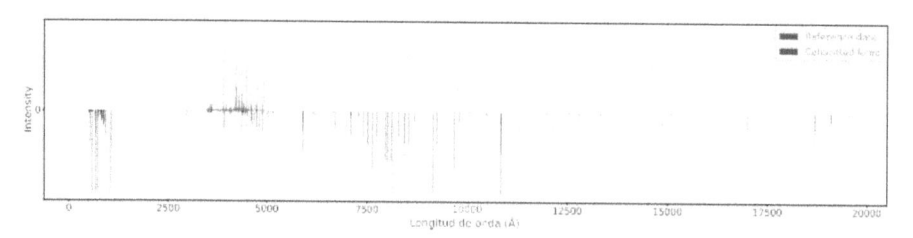

Fig. 7. A comparison is presented between the extracted spectrum of a He-Ne-Ar lamp from Complejo Astronomico el Leoncito (center up in black) and the reference spectrum constructed from NIST data (center down in blue). (Color figure online)

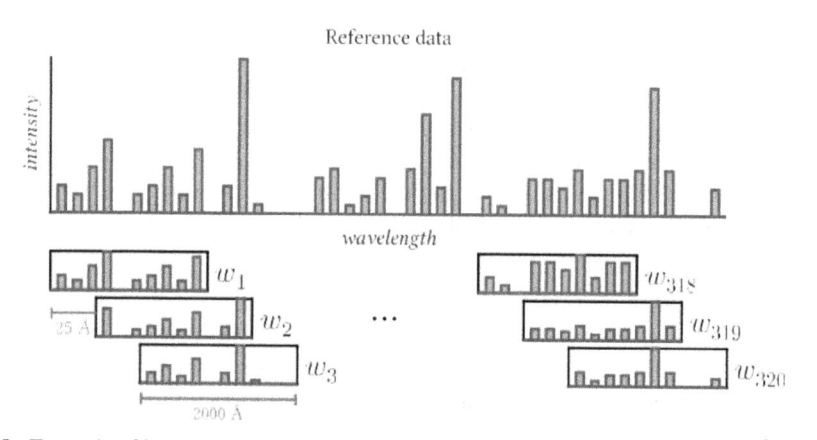

Fig. 8. Example of how we segment reference data. Note that each window starts 25 Å after its predecessor and each window generates data within 2000 Å of its start.

3.4 Identification of the Best Segment

From each alignment, we compute the DTW distance, which represents how much the observed function had to be deformed to fit the reference data.

From the experiments carried out, the window with the lowest distance has a high probability of corresponding to the window that best matches the wavelengths that must be aligned. The above is true as long as the lamp elements are well indicated, since otherwise erroneous reference data usually have a great impact on the quality of the alignment.

3.5 Evaluation Metric

Given 2 calibrations of the same lamp, one manual performed by an expert in the field and another. It is possible to evaluate the quality of the second calibration using the "Intersection over Union" (IoU) metric. IoU is calculated by taking the wavelength range of both calibrations. Then, it is determined how many values coincide to calculate the Intersection, and the number of wavelengths they occupy between the 2 is determined to calculate the Union (counting the values repeated only once). Then *Intersection/Union* is calculated. If IoU has a value close to one, it means that the ranges coincide to a greater or total extent, therefore it is likely that the calibration being evaluated is good. If IoU has a value close to 0, it means that the ranges have little or no overlap, implying that the calibration being evaluated is bad.

3.6 Experiments

To test the veracity of the lamp calibration method, we experimented with 32 samples of He-Ne-Ar lamp spectra using a Boller and Chivens spectrograph attached to the 2.15m Jorge Sahade telescope at the *Complejo Astronómico El Leoncito* (CASLEO), San Juan, Argentina and their manually calibrated versions.

For each comparison lamp function sample we apply a series of steps:

1. We obtain the reference data corresponding to the chemical elements used by the lamp and divide them into windows (each one has a length of 2,000 Å and there are 25 Å between the beginnings of each window)
2. Occasionally the reference data have gaps of various wavelengths between which they do not report any intensity data. It may happen that there are windows that have fallen right into one of these gaps and therefore have been left empty. In these cases it is definitive that the wavelengths of the window will not correspond to those of the lamp function, so if they exist, they are discarded.
3. We apply DTW on each of the windows in combination with the data of the lamp that we are calibrating and we are left with the one that finally shows the lowest DTW Distance.
4. From the chosen window we calculate the "IoU" (Intersection over Union) metric to identify how well the automatic calibration was performed. As a reliable calibration we use the manual calibration that corresponds to the lamp being calibrated.

During the executions there were certain parameters that were varied to find the best configuration for carrying out calibrations.

1. The origin of the reference data: We tested the 2 aforementioned data sources NIST and LIBS. Given these sources, 3 sets of data were recovered that can potentially

be used as reference data: first, NIST data with the peaks of the chemical elements He, Ar and Ne. Second, LIBS data with the same chemical elements where values intermediate to the peaks are inferred, with a resolution of 100 Å (LIBS100). Third, the same as the second but with a resolution of 260 Å (LIBS260). The variation in resolution between set 2 and 3 was purposely sought to study which granularity level is most desirable during the calibration process. NIST data only has information on interest peaks.

2. The second parameter that we were interested in varying was whether or not an extraction of peaks from the reference data (**PR**) was performed. In the case of the NIST data, there is not much difference since from the beginning these are peak data. However, in LIBS data the extraction of peaks involves a strong reduction in dimensionality. It is also important to note that when performing peak extraction on the values obtained from LIBS, the resulting values will not necessarily coincide with the peak values shown by NIST, mainly due to the specialized calculations with which LIBS infers the values between data peaks.

3. Third, we parameterize the peak extraction in the comparison lamp (**PL**) function. Performing this extraction always involves a strong dimensionality reduction in the lamp data, which potentially improves the execution times of the algorithm. It is interesting to distinguish whether this reduction also represents a reduction in the quality of the calibration.

4. Fourthly, it is parameterized whether the intensity values contained in each window are normalized or not (**WN**). By default a normalization is performed on all the reference data, however this presents a situation in which when dividing the reference data into windows, these can be left with very small data, which made sense when a total normalization is performed, but they lose it by separating the higher reference values in another window. The fact that all the values are very small and have very similar values can be detrimental to the alignment algorithm. To avoid these situations and allow the windows to have very similar data, a normalization is carried out in each individual window, in this way the small differences between the data of a window can be seen despite the fact that in the entire data reference only seems like a small disturbance. After all, it is quite possible that the signal from the comparison lamp corresponds to peaks that appear small in the overall image.

5. Finally, zero padding (**ZP**). This option consists of adding intensity values of 0 to the reference data when series of wavelengths are found among which no intensity data were found. The option makes more or less sense depending on the configuration that is being used to obtain data since for high resolution data practically no data aggregation is performed.

Considering all the options, we evaluated the method with 3 data sources, PR, PL, WN and ZP. Therefore, we have to consider a total of 48 possible configurations. The processing of the 32 samples was carried out with each of these and on each occasion the metrics of interest were collected. Table 1 shows the top 4 calibrations for each reference data source. The complete testing table can be found in the Annex A.

Table 1. Best parameter configurations found for the task of calibrating comparison lamp functions. **(a)** Top 4 configurations found when using NIST reference data. **(b)** Top 4 configurations found when using LIBS reference data with a resolution of 100 Å. **(c)** Best 4 configurations found when using LIBS reference data with a resolution of 260 Å.

Using NIST reference data

PR	PL	WN	ZP	Count of Windows	(AVG) IoU	(STD) IoU
✗	✗	✓	✓	1620	**0.93**	**0.01**
✗	✗	✗	✗	1039	0.84	0.03
✓	✗	✗	✗	**846**	0.80	0.02
✓	✗	✓	✓	846	0.79	0.02

(a)

Using LIBS100 reference data

PR	PL	WN	ZP	Count of Windows	(AVG) IoU	(STD) IoU
✓	✗	✓	✓	**763**	**0.79**	0.07
✓	✓	✓	✗	745	0.65	0.26
✓	✗	✓	✗	745	0.39	0.02
✓	✗	✗	✗	745	0.37	**0.00**

(b)

Using LIBS260 reference data

PR	PL	WN	ZP	Count of Windows	(AVG) IoU	(STD) IoU
✓	✓	✓	✗	**763**	**0.72**	0.04
✓	✗	✓	✓	763	0.69	**0.00**
✓	✗	✓	✗	763	0.59	0.00
✓	✓	✓	✓	763	0.33	0.25

(c)

3.7 Analysis of Results

As can be seen in the results of Table 1, the average performance of the calibrations carried out tends to be better the lower the resolution of the data used. Since the best performance is found in the Table 1a in its first row and as we already know, these NIST data are composed only of the intensities and wavelengths of the peaks belonging to the electromagnetic spectra of the chemical elements interest. When high resolution data was used (LIBS100 and LIBS200) the PR option was always activated so the results seem to favor those configurations that use small amounts of reference data. In the NIST configurations, activating PR worsened the quality of the calibration, but this was natural since this option further reduced the already limited information available when using NIST reference data.

Detecting spikes in the lamp data using the PL option does not appear to have been a priority in most cases. It seems to have contributed a little when dealing with high resolution data where the PR option was also enabled (Table 1c, rows 1 and 4. Table 1b, row 2). However, in this case the standard deviation was particularly high, which leads us to think that this combination of options also brings with it an increase in how variable the quality of the calibration is.

The WN option was activated for most of the best configurations. Although, when working with NIST data (Table 1a) this option was not used in the second and third best configurations. However, the best performance of all did require this option to be active in conjunction with ZP, the best configuration excluding this did not have any option activated. The use of WN in conjunction with ZP meant an improvement of ≈ 0.09 in the average IoU.

From ZP it was observed that in cases in which there was little data, the use of this option together with WN represented an improvement compared to the cases in which it was used alone (Table 1a and b). This can be explained by thinking about how the normalization of data segments is carried out with WN. To normalize a window we take the minimum and maximum of the data array that we want to normalize and with these two values we transform the data to values between 0 and 1. When we only have information on the most relevant peaks (as in NIST) then although the maximum information is correct, as a minimum the value of the smallest peak is used. Then, when normalizing and calibrating, the implicit information that there should be more values between the peaks is lost. ZP improves this situation since by also filling the data gaps with zeros it forces the minimum value of those windows where data is missing to be yes or yes 0. This allows the normalization to maintain a correct proportion of the real height handled by the data.

On the contrary, in situations with a larger amount of data (Table 1c) the use of the ZP option in conjunction with WN can even be harmful. In those situations in which PR is not used, nothing happens since by not decreasing the reference data, the high resolution of the data is maintained and by not finding gaps in wavelengths without intensities, ZP has no effect. However, when PR is activated, a single selection of the reference data peaks is made, which generates gaps between the wavelengths that are subsequently filled with zeros using the ZP option. The underlying problem is the elimination of the information about the decays between peaks to replace them directly with zeros. Faced with such a loss of information, a decrease in the quality of the calibrations is natural.

4 Conclusions

As final conclusions, we draw that the methodology of segmenting the reference data into information windows and applying DTW in each of them to obtain the final calibration proved to be a viable method to calibrate comparison lamps. In the best of cases, a configuration was achieved that manages to calibrate 93% of a comparison lamp in the correct range of wavelengths. This is not the optimal result for use in production, but it gives us a good starting point from which to iterate until we obtain a version that achieves calibrations with the precision that the problem demands.

Source of reference	PT	PE	WN	ZP	Count of Windows	(AVG) IoU	(STD) IoU
NIST	✗	✗	✓	✓	1620	0.93	0.01
NIST	✗	✗	✗	✗	1039	0.84	0.03
NIST	✓	✗	✗	✗	846	0.80	0.02
NIST	✓	✗	✓	✓	846	0.79	0.02
LIBS100	✓	✗	✓	✓	745	0.79	0.07
NIST	✗	✗	✗	✓	1620	0.78	0.00
NIST	✓	✓	✓	✗	846	0.78	0.06
NIST	✓	✗	✗	✓	846	0.74	0.03
LIBS260	✓	✓	✓	✗	763	0.72	0.04
LIBS260	✓	✗	✓	✓	763	0.69	0.00
LIBS100	✓	✓	✓	✗	745	0.65	0.26
NIST	✗	✗	✓	✗	1039	0.62	0.07
LIBS260	✓	✗	✓	✗	763	0.59	0.00
NIST	✓	✓	✓	✓	846	0.56	0.04
NIST	✓	✓	✗	✗	846	0.50	0.17
NIST	✓	✗	✓	✗	846	0.44	0.07
NIST	✗	✓	✓	✗	1039	0.43	0.03
LIBS100	✓	✗	✓	✗	745	0.39	0.02
LIBS100	✓	✗	✗	✗	745	0.37	0.00
LIBS260	✓	✓	✓	✓	763	0.33	0.25
NIST	✗	✓	✓	✓	1620	0.26	0.12
LIBS100	✓	✓	✗	✗	745	0.20	0.13
LIBS100	✓	✓	✓	✓	745	0.08	0.11
LIBS100	✗	✗	✓	✓	799	0.01	0.00
LIBS100	✗	✗	✓	✗	799	0.01	0.00
LIBS100	✗	✓	✓	✓	799	0.00	0.00
LIBS100	✗	✓	✓	✗	799	0.00	0.00
LIBS100	✗	✓	✗	✓	799	0.00	0.00
LIBS100	✗	✓	✗	✗	799	0.00	0.00
LIBS100	✓	✗	✗	✓	745	0.00	0.00
LIBS100	✗	✗	✗	✓	799	0.00	0.00
LIBS100	✗	✗	✗	✗	799	0.00	0.00
LIBS100	✓	✓	✗	✓	745	0.00	0.00
LIBS260	✗	✗	✓	✓	781	0.00	0.00
LIBS260	✗	✗	✓	✗	781	0.00	0.00
LIBS260	✗	✓	✓	✓	781	0.00	0.00
LIBS260	✗	✓	✓	✗	781	0.00	0.00
LIBS260	✓	✗	✗	✗	763	0.00	0.00
LIBS260	✗	✓	✗	✓	781	0.00	0.00
LIBS260	✗	✓	✗	✗	781	0.00	0.00
LIBS260	✗	✗	✗	✓	781	0.00	0.00
LIBS260	✗	✗	✗	✗	781	0.00	0.00
LIBS260	✓	✓	✗	✗	763	0.00	0.00
LIBS260	✓	✗	✗	✓	763	0.00	0.00
LIBS260	✓	✓	✗	✓	763	0.00	0.00
NIST	✗	✓	✗	✓	1620	0.00	0.00
NIST	✗	✓	✗	✗	1039	0.00	0.00
NIST	✓	✓	✗	✓	846	0.00	0.00

A series of behaviors were also observed that can guide us when perfecting the methodology during future research. In particular:

- It is important to keep the reference data that will be used simple and with only the relevant information (preferably the peak values). This, in addition to improving calibration, also improves processing times by using less data during alignment.
- It is not necessary to reduce the dimensionality of the comparison lamp function data, in most cases this is detrimental and when it is not it causes the results to vary too much to be reliable.
- Filling with zeros in the wavelength spaces turns out to be a behavior that favors the quality of the calibration, it allows maintaining the logic of the sequences of the data that are being compared, avoiding assuming that there are only straight lines between 2 peaks. Furthermore, it was observed that filling zeros increases performance in those cases in which the windows to be calibrated are also being normalized. As long as the data source has low resolution, otherwise its use is counterproductive.

5 Future Work

As already said, the results obtained are not yet optimal for use in production, so in the future it is planned to perfect the methodology to achieve greater precision in the quality of the comparison lamp calibrations that are carried out.

In this set of experiments, the test data set was not ideal. Although there were several rows of lamps, they were all made up of the same proportions of chemical elements (98% Helium, 1% Argon, 1% Neon). It is especially difficult to find manual calibrations already performed for this type of data, so we started from them to see if the processing method was viable. The results were positive so we will try to acquire more variety of hand-calibrated lamps for future works.

Another alternative to deal with the shortage of testing data could be to generate artificially calibrated lamps. Obviously these lamps would not correspond to real observations but rather would be generated artificially based on the reference data available. The biggest challenges for this are the domain knowledge needed to understand what lamps are like, what their typical values are, and what rules they must follow to make sense. Furthermore, given the characteristics of the analysis where it works with old plates. It is necessary to simulate the defects and deformations due to prolonged storage that real plates usually bring with them, among other things.

Both perspectives for obtaining more testing data are considered viable, so they will be investigated for application in future research.

Acknowledgements. YJA thank the financial support from the *Universidad Nacional de La Plata*. (Proyectos I+D 11/G193 y EG001), Argentina.

A Annex: Complete Table of Tested Calibration Settings

References

1. DASCH: Dasch: Digital access to a sky century @ harvard a new look at the temporal universe (2017). http://dasch.rc.fas.harvard.edu/project.php
2. Dawson, B.H., Shapley, H.: Bright Nova. Harv. College Obs. Announc. Card **637**, 2 (1942)
3. Giorgino, T.: Computing and visualizing dynamic time warping alignments in R: The dtw package. J. Stat. Softw. **31**(7), 1–24 (2009). https://doi.org/10.18637/jss.v031.i07. https://www.jstatsoft.org/index.php/jss/article/view/v031i07
4. Kramida, A., Olsen, K., Ralchenko, Y.: NIST LIBS database. National Institute of Standards and Technology, US Department of Commerce (2019)
5. Mandel, H., Birkle, K., Demleitner, M., Heidelberg, L.: HDAP – Heidelberg digitized astronomical plates. VO resource provided by the GAVO Data Center (2007). https://doi.org/10.21938/haTEMZmoaCTEK6XZvGU.fQ. http://dc.zah.uni-heidelberg.de/lswscans/res/positions/q/info
6. Meilán, N.S.: Recuperacion del patrimonio observacional histórico. Bachelor's thesis, Universidad Nacional de La Plata (2018)
7. Ponte Ahon, S.A.: Digitalización de placas astronómicas antiguas. Bachelor's thesis, Universidad Nacional de La Plata (2023)
8. Ralchenko, Y.: Nist atomic spectra database. Memorie della Societ'a Astronomica Italiana Supplement, v. 8, p. 96 (2005)
9. Raouph, M., Schrimpf, A., Kroll, P.: Prospect of plate archive photometric calibration by GAIA SED fluxes (2023)
10. ReTrhOH: Retrhoh — recuperación del trabajo observacional histórico (2019). https://retroh.fcaglp.unlp.edu.ar/astronomia/
11. Ronchetti, F., et al.: Software inteligente para la digitalización de placas espectroscópicas. In: XXVIII Congreso Argentino de Ciencias de la Computación (CACIC)(La Rioja, 3 al 6 de octubre de 2022) (2023)
12. Tormene, P., Giorgino, T., Quaglini, S., Stefanelli, M.: Matching incomplete time series with dynamic time warping: an algorithm and an application to post-stroke rehabilitation. Artif. Intell. Med. **45**(1), 11–34 (2009). https://doi.org/10.1016/j.artmed.2008.11.007. https://www.sciencedirect.com/science/article/pii/S0933365708001772
13. Tuvikene, T.: Pyplate repository (2022). https://github.com/astrotuvi/pyplate
14. UAI: Nro. b3 safeguarding the information in photographic plates (2000). https://www.iau.org/static/publications/ib88.pdf

An Empirical Method for Processing I/O Traces to Analyze the Performance of DL Applications

Edixon Parraga[1]($\boxtimes$), Betzabeth Leon[1], Sandra Mendez[2], Dolores Rexachs[1], Remo Suppi[1], and Emilio Luque[1]

[1] Computer Architecture and Operating Systems Department, Universitat Autònoma de Barcelona, Campus UAB, Edifici Q, 08193 Bellaterra, Barcelona, Spain
{edixon.parraga,betzabeth.leon,dolores.rexachs,
remo.suppi,emilio.luque}@uab.es

[2] Computer Sciences Department, Barcelona Supercomputing Center (BSC), Barcelona 08034, Spain
sandra.mendez@bsc.es
https://webs.uab.cat/hpc4eas/

Abstract. The exponential growth of data handled by Deep Learning (DL) applications has led to an unprecedented demand for computational resources, necessitating their execution on High Performance Computing (HPC) systems. However, understanding and optimizing Input/Output (I/O) of the DL applications can be challenging due to the complexity and scale of DL workloads and the heterogeneous nature of I/O operations. This paper addresses this issue by proposing an I/O traces processing method that simplifies the generation of reports on global I/O patterns and performance to aid in I/O performance analysis. Our approach focuses on understanding the temporal and spatial distributions of I/O operations and related with the behavior at I/O system level. The proposed method enables us to synthesize and extract key information from the reports generated by tools such as Darshan tool and the seff command. These reports offer a detailed view of I/O performance, providing a set of metrics that deepen our understanding of the I/O behavior of DL applications.

Keywords: DL · I/O Analysis · HPC · I/O behavior patterns

1 Introduction

Deep learning (DL) is one of the most popular computational approaches in the field of machine learning (ML) and has been shown to improve performance in

---

This research has been supported by the Agencia Estatal de Investigación (AEI), Spain, and the Fondo Europeo de Desarrollo Regional (FEDER) UE, under contract PID2020-112496GB-I00 and partially funded by the Fundacion Escuelas Universitarias Gimbernat (EUG). The authors thankfully acknowledge RES resources provided by CESGA in FinisTerrae III to RES-DATA-2022-1-0014.

© The Author(s), under exclusive license to Springer Nature Switzerland AG 2025
M. Naiouf et al. (Eds.): JCC-BD&ET 2024, CCIS 2189, pp. 74–90, 2025.
https://doi.org/10.1007/978-3-031-70807-7_6

areas such as natural language processing [1], computer vision [2], and computational biology [3]. However, the exponential growth of data handled by DL applications has led to an unprecedented demand for computational resources, necessitating their execution on High Performance Computing (HPC) systems [4].

A critical and often challenging aspect of running DL applications in HPC systems is efficiently managing file Input/Output (I/O) operations. These operations are essential for loading data and storing results. Optimal handling of these operations is important, especially when the application needs to deal with datasets containing thousands of samples (i.e., thousands of small images) processed by the parallel file system of the HPC systems. These operations present I/O patterns that can become bottlenecks, limiting the overall performance of DL applications [5].

To understand the impact of I/O on the performance of DL applications, it is necessary to comprehend their I/O behavior in HPC systems. However, due to the complexity of the I/O software stack for these applications, there is a need for a method to process and depict I/O metrics and patterns in a structured way that can guide users in the analysis of I/O performance.

In this context, we propose an I/O traces processing method that simplifies the generation of reports about global I/O patterns and performance by utilizing specific outputs from monitoring and profiling tools. Our approach focuses on analyzing temporal and spatial I/O patterns and their distribution on the parallel filesystem, which can be correlated with the obtained performance metrics. As the authors note in [6], analyzing and understanding an application's I/O access patterns provides key insights into how an application's I/O behavior affects its performance on different systems. This allows us to understand the I/O behavior on the HPC system and identify and data access and distribution patterns that minimize their impact on application I/O performance. Therefore, our paper aims to contribute in the following ways:

- Introducing an I/O trace processing method that extracts and synthesizes critical information from outputs generated by I/O monitoring and profiling tools in HPC systems.
- Providing insights to aid in the identification of potential I/O bottlenecks within the output of compute nodes and/or storage nodes.

The structure of this article is presented as follows: Sect. 2 highlights the usefulness of understanding I/O patterns by using profiling, monitoring and tracing tools and it also provides a review of related works. Section 3 presents our I/O trace processing method step by step. Section 4 applies our approach to a case study. Finally, in Sect. 5, we present our conclusions and future work.

2 Background

In this section we present a brief review on the importance of using I/O tracing, monitoring, and profiling tools to understand the I/O behavior of DL applications. Furthermore, we review some works related to our approach.

2.1 Analysis from Data Traces and Monitoring Tools

Understanding the I/O behavior in DL applications requires a meticulous and structured approach to capturing and analyzing meaningful data. This section breaks down into two main components that our approach needs: 1) Patterns of Data Access in Files and 2) Profiling, Tracing and Monitoring tools to capture these access patterns at the different levels of the I/O software stack.

Patterns of Data Access in Files: According to the authors of [7], data access characteristics in DL workloads differ markedly from traditional workloads, with unique patterns in memory and random accesses to large files. Tracing and analysing I/O operations provide a detailed insight into data access and allow measurement of the impact on application performance. This is essential for understanding and optimizing the I/O of the DL application. Traces collect important information about resource usage and I/O operations behavior. These traces help in identifying bottlenecks and potential issues, and they are essential for validating and reproducing experiments.

Profiling, Tracing and Monitoring Tools: To analyze the I/O behavior of the DL applications addressed in this study, the Darshan tool [6] has been selected. Darshan is a tool designed to investigate the I/O performance of HPC applications at large scale. Darshan is composed of different modules that allow it to capture the I/O operations at different levels. Furthermore, Darshan can be used for profiling, tracing, or monitoring of the file system operations, but on the client side. That is, this tool is deployed in the same environment where the DL application is running, providing direct, real-time monitoring of its I/O performance.

Darshan logs contains detailed information about I/O operations, timing, performance, I/O counters and so on. Synthesizing and organizing this extensive information is essential for practical analysis, which requires a processing method that condenses and clarifies the data. Using as input Darshan's logs, our proposed method simplifies the information and facilitates the visualization of the results of I/O behavior.

2.2 Related Work

The behavior of I/O operations in DL is very important, being the focus of studies such as [8] and [9], which examine some existing analytical tools and compare some DL models. In [10], the authors use DXT Explorer to optimize these I/O operations. The DLIO tool, introduced in [11], replicates I/O behavior in scientific DL workloads, which is helpful for DL performance evaluation and improvement. Additionally, the research in [12] focuses on the evolution of I/O evaluation, while the research in [13] deals with the emulation of I/O behavior in scientific workflows. These studies highlight the influence of new workloads on HPC systems and the need to understand data transfer between modules.

Our work distinguishes itself by developing a method that processes I/O traces, providing a detailed analysis of I/O behavior and identifying potential improvements, thus contributing to research on I/O performance in DL applications.

3 Proposed Trace Analysis Method

This section describes our proposed method with the essential steps to perform the I/O trace processing of a DL application. The objective is to accurately obtain the required information, which is fundamental for analyzing the DL application I/O. The proposed method is composed by three main stages: Input, Processing and Output.

3.1 Stage 1: Input

In this stage, we set the right environment for the different tools used to capture the information related to I/O activities.

I/O Data Acquisition Tool Deployment: In the initial phase, we enable the DXT Darshan [6] module to obtain traces of the I/O operations at POSIX-IO level. This detail is needed to identify and model the I/O spatial and temporal pattern of the application. Darshan is loaded by using the LD_PRELOAD environment variable and this is exported in job submission script before the command to run the application in the HPC system. If the application runs without any problems, a Darshan log file is generated per each job identifier. Furthermore, the information related with counters and performance is provided by Darshan parser and perf.

As we deploy this tool in HPC systems, we implement scripts to take the job identifier of each execution of the application to relate with each Darshan log generated. This allows us to process several Darshan logs at the same time. It is useful when users need to analyze scaling issues.

Additionally, to see CPU and memory usage for a job and evaluate if the I/O is having impact on these resources we use the seff command output of the Slurm Workload Manager.

3.2 Stage 2. Processing

In this stage, we initiate monitoring of the application by submitting the job to the SLURM queuing system (using the command sbatch script.sh), ensuring that the monitoring configuration established in the previous stage is fully active. To guarantee the accuracy and reliability of the data obtained, it is essential to perform each experiment several times. During this process, we verify that our Job is running correctly (command squeue) to confirm that the monitoring system is working correctly and that the binary file .darshan is being generated completely and consistently. This step ensures the integrity of the collected data and provides a basis for subsequent analysis.

Data Preparation: It involves organizing and preparing the collected data for analysis, including cleaning, filtering out irrelevant data, and transforming it into a format suitable for analysis. The data selected from the different reports are presented in Table 1. This table describes the relevant information extracted from the original Seff and Darshan reports, which were obtained directly without any additional processing. Additionally, it details the specific element extracted from each report and explains its usefulness in the context of our work.

Table 1. Data selected for the traces I/O analysis

Report	Information provided	Element extracted from the report	Utility
Seff	Job Metadata	Job ID, cluster, user/group, status	Allows you to track and audit resource usage. Provides context for the execution of the job
	CPU and Memory Efficiency	CPU used, CPU efficiency, Core-walltime, memory efficiency.	Identifies resource utilization efficiency, helping to optimize CPU and memory allocation
	Computing and Memory Resources	Nodes, cores per node, job clock time, memory used.	Provides data on the processing capacity used. Helps evaluate execution time in relation to allocated resources
Darshan DXT Report	General Job Metadata and File System	Job ID, nprocs, runtime, file id, fs type, Lustre stripe size, Lustre OST obdidx.	Allows mapping I/O operations to specific resources
	POSIX Module Data	File name, rank, hostname, write count, read count, Lustre stripe count, Module.	Provides data on the volume of I/O generated. Helps to understand how the workload is distributed
	Temporal and Spatial Tracking Metrics	Wt/Rd, Segment, Offset, Length, Start(s), End(s), OST.	Provides specific details about I/O operations, essential for identifying and resolving bottlenecks
Darshan Parser Report	Identification and Context of Work	Job ID, nprocs, run time, record_id, Module, fs type, file name	Defines the context and uniqueness of the analyzed work
	POSIX Performance Metrics	POSIX READS/WRITE, POSIX BYTES, POSIX MAX BYTE, POSIX SIZE, POSIX F READ, POSIX F WRITE	Provides a detailed view of the performance of POSIX operations
	Performance Analysis in LUSTRE	Number of OSTs, stripe parameters, POSIX time	Provides a detailed understanding of how I/O resources are configured and used
Darshan Perf Report	General Job Metadata	Job ID, nprocs, run time, total bytes	Helps understand the scale of operations and I/O intensity
	I/O Performance Metrics	I/O timing for unique files (seconds), I/O timing for shared files (seconds), Aggregate performance	Allows you to evaluate the impact of concurrent operations on overall performance

Report Generation and Analysis: the collected traces are processed to generate reports summarizing the performance and efficiency of I/O operations. This includes descriptive statistics, visualizations, and the detection of initial patterns.

A Report: Utilized Resources by the Job. The utilized resources per job can be obtained by the `seff` command line. `seff`'s output presents mainly information related to the memory utilization and CPU efficiency for completed jobs. The data obtained includes the following:

1. Left Column - Job Metadata: `Job ID`: This is the unique identifier assigned to the job by the cluster management system. `Cluster`: The specific cluster in which the job was run that is relevant to understanding the operational context. `User/Group`: The identity of the user or user group that submitted the job is critical for auditing and tracking resource utilization. `State`: The final status of the job (e.g., successfully completed, failed, canceled), providing an immediate view of the job's outcome without going into specific performance details.
2. Center Column - CPU and Memory Usage Efficiency Metrics: `CPU Utilized`: is the total CPU time the job has utilized. `CPU Efficiency`: A measure that reflects how effectively the allocated CPU time has been used, calculated as the ratio of CPU time to total available CPU time. `Core-walltime`: The product of the wall time and the number of cores, providing a composite measure of processing time consumed. `Memory Efficiency`: Similar to CPU efficiency, this metric evaluates the efficiency of memory usage allocated to the job.
3. Right Column - Compute and Memory Resources Used: `Nodes`: The total number of compute nodes assigned to the job. `Cores per node`: The number of processor cores available per node, which is essential to understanding the computing capacity per node. `Job Wall-clock time`: The total time from start to completion. `Memory Utilized`: The memory used during the job execution.

Figure 1 shows in orange the data from the original report and in sky-blue the data extracted for the performance analysis in the next steps of our method. The three main columns represent the different categories of data to be extracted for the report.

B Darshan Log Files: We run our scripts, which process the log files generated by Darshan while monitoring the application, to generate the `DXT, PARSER`, and `PERF` reports. These reports are then used to apply our method for analyzing the I/O performance of the application. Our method focuses on the I/O done by the application on the datasets.

Darshan DXT Report Processing: This processing consists of extracting and organizing the relevant information from the DXT reports. Figure 2 shows the main information into three columns, representing different strata of metadata and metrics:

1. General Metadata of the Job and the File System (Left Column): `Job ID, nprocs, run time`: These attributes define the operational context of the

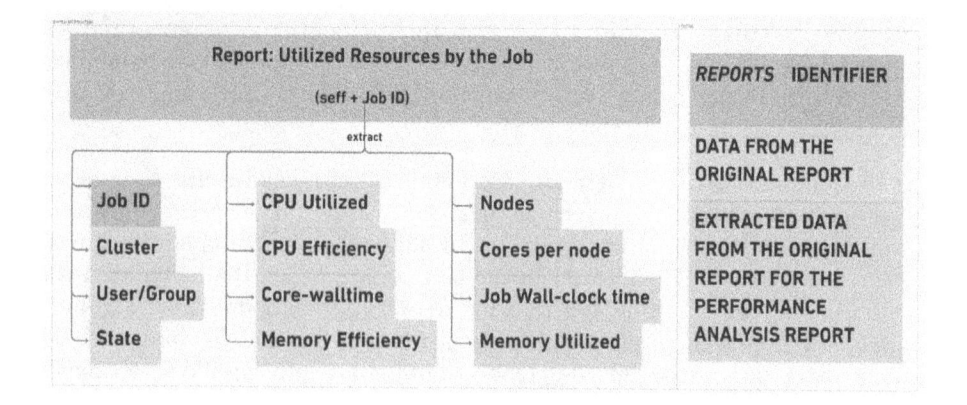

```
                    Report: Utilized Resources by the Job
                              (seff + Job ID)                        REPORTS IDENTIFIER
                                    extract
                                                                     DATA FROM THE
                                                                     ORIGINAL REPORT
    Job ID          CPU Utilized          Nodes
                                                                     EXTRACTED DATA
    Cluster         CPU Efficiency        Cores per node             FROM THE ORIGINAL
                                                                     REPORT FOR THE
    User/Group      Core-walltime         Job Wall-clock time        PERFORMANCE
                                                                     ANALYSIS REPORT
    State           Memory Efficiency     Memory Utilized
```

Fig. 1. Utilized resources by the Job

computational job, including the unique job identification, process concurrency, and duration. `file id, fs type`: These are the identifiers that specify the file and the type of file system used, essential for mapping I/O operations to a specific storage resource. `Lustre stripe size, Lustre OST obdidx`: These are parameters related to the configuration of LUSTRE, a parallel file system, where the 'stripe size' and 'OST obdidx' are critical to understanding data distribution and storage location.

2. POSIX DXT Module Data (Central Column): `file name, rank, hostname`: These fields directly reference the file and execution context, including the rank of the MPI process and the compute node. `write count, read count`: Quantitative metrics of read and write operations provide insight into the volume of I/O that a specific job generates. `Lustre stripe count, Module`: Lustre's stripe count and the identification of the Darshan module used indicate the parallelization configuration and the active tracking subsystem.

3. Temporal and Spatial Tracking Metrics (Right Column): `Wt/Rd, Segment, Offset, Length, Start(s), End(s), OST`: These parameters outline the I/O profile at a granular level, detailing whether the operations are read or write, the specific location within of the file (segment and offset), the amount of data involved (length), and the timing of the operations (start and end), as well as the Object Storage Targets (OSTs).

Darshan Parser Report Processing: `darshan parser` command provides detailed set of I/O performance metrics and counters for a specific job executed in a HPC system. Below, we summarize the main information extracted from the report to be used by our method:

1. Job Identification and Context: attributes such as Job ID, number of processes (nprocs), run time (run time), record identifier (record_id), module used (Module), file system type (fs type), and file name (file name) are essential to characterize the operational context and uniqueness of the analyzed job.

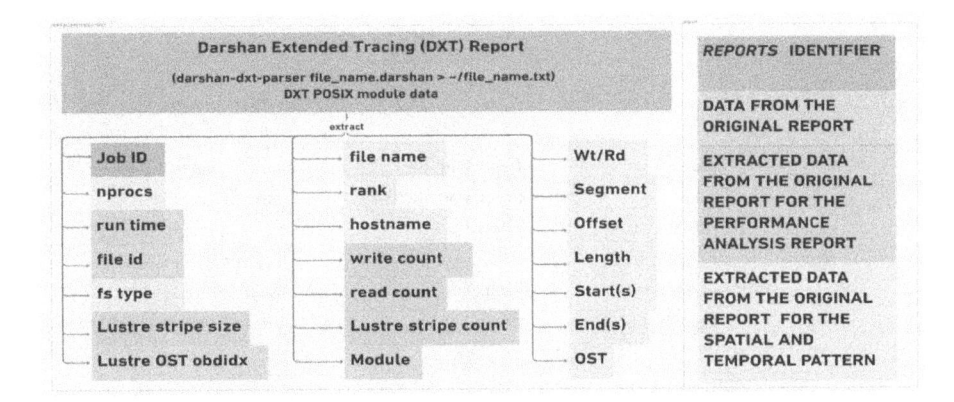

Fig. 2. Main I/O data extracted from DXT report

2. POSIX Performance Metrics: metrics related to the POSIX module, offering an overview of the number and volume of read and write operations (POSIX READS/WRITE), byte access (POSIX BYTES and POSIX MAX BYTE), and the sizes of the I/O operations (POSIX SIZE * - *). The specific entries for reading and writing files (POSIX F READ and POSIX F WRITE) offer a more focused look at individual file operations.
3. Performance Analysis in LUSTRE: The metrics associated with the LUSTRE file system provide detailed insight into the configuration and performance of a parallel file system. Metrics include the number of Object Storage Targets (OSTs), which directly influence the capacity and speed of I/O operations, and stripe parameters (size and width), which determine how data are distributed and accessed across multiple OSTs. Additionally, the time associated with metadata operations (POSIX F META TIME) is important to understanding the metadata management overhead.

In Fig. 3, it can be observed in sky-blue the selected information from the original report to be used to synthesize new analytical reports. These reports are important for understanding the performance of a specific job and for helping to identify potential I/O bottlenecks and optimization points.

Darshan Perf Report Processing: `darshan perf` command provides I/O performance metrics for shared and independent files accessed by application. As our focus is on performance evaluation, we extract the following data:

1. General Job Metadata (Left Column): `Job ID`: It is the identifier of the job, providing a unique reference to the set of supercomputing tasks in question. `nprocs`: Indicates the number of parallel processes running, which is critical to understanding the scale of parallelism involved. `run time`: Represents the duration of the supercomputing work, which is essential for evaluating time efficiency. `total bytes`: Quantifies the total volume of data handled, providing a metric of the I/O intensity of the work.

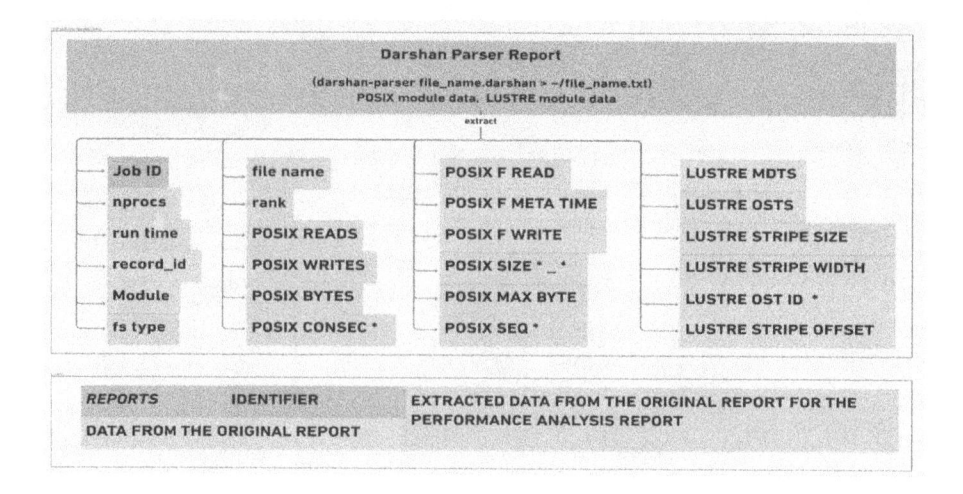

Fig. 3. Darshan Parser Report

2. I/O Performance Metrics (Right Column): `I/O timing for unique files (seconds)`: Reflects the I/O time spent on unique files, essential for discerning file system performance when processes access their own files without sharing. `I/O timing for shared files (seconds)`: Measures the I/O timing for files shared between processes, which can indicate how concurrent operations affect performance. **Aggregate performance**: Displays aggregate performance metrics based on the slowest time recorded, which could identify the worst-case performance among all processes and provide insights for optimizations.

Figure 4 presents data selected in sky-blue from the original report, which are used in performance analysis in our method.

Performance Analysis Report: Based on the previous reports obtained in the base processing, our automated method processes the files and creates an overall report consolidating the previous reports, using the JobID as an index.

This global report summarizes the calculation of the information relevant to our analysis. Among the relevant data and its calculations are the following: Total Metadata Operations, Total Data Access Operations, Data Access Operations by Nodes, Data Access Process Operations, I/O Operations per second (IOPs), Bytes per Node, Bytes per Process, Total Operations. Figure 5 depicts the different original reports and data selected from them to be used by our method. The selected data allows us to analyze the I/O performance of the DL application.

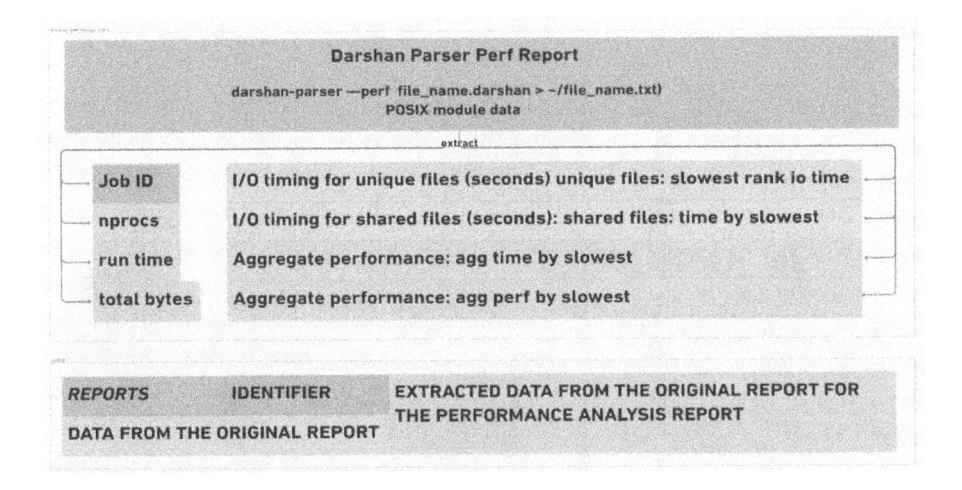

Fig. 4. Perf report

3.3 Stage 3. Output

Taking all the information selected from the original reports two main outputs are available for the user: 1) Spatial and Temporal Pattern Analysis and 2) Performance Analysis.

Spatial and Temporal Pattern Analysis: An analysis of the I/O operations from DXT reports is carried out to identify temporal and spatial patterns. Users can plot these patterns and analyze the behavior of I/O events based on their order or timestamp. This analysis enables users to identify if there is a serialization of I/O operations by examining the sequence and timing of these operations. If a large number of I/O operations occur in a sequential manner without overlapping, this indicates serialization.

Furthermore, by comparing these patterns against expected parallel behavior, users can determine if the serialization is caused by the underlying I/O system's inability to handle parallel requests efficiently or by the I/O pattern generated by the application itself. Spatial and temporal access patterns are influenced by both the file type and the file system. While the file type dictates the logical organization of data, the file system's design and performance characteristics can affect how efficiently these patterns are handled, especially in parallel file systems.

Additionally, these patterns can be used to extrapolate the count of I/O operations and total bytes for different numbers of processes and I/O workloads, providing valuable insights for scaling and optimization.

Performance Analysis: Finally, the performance of the I/O operations is evaluated based on the previous analysis, as shown in Fig. 5. This step may involve

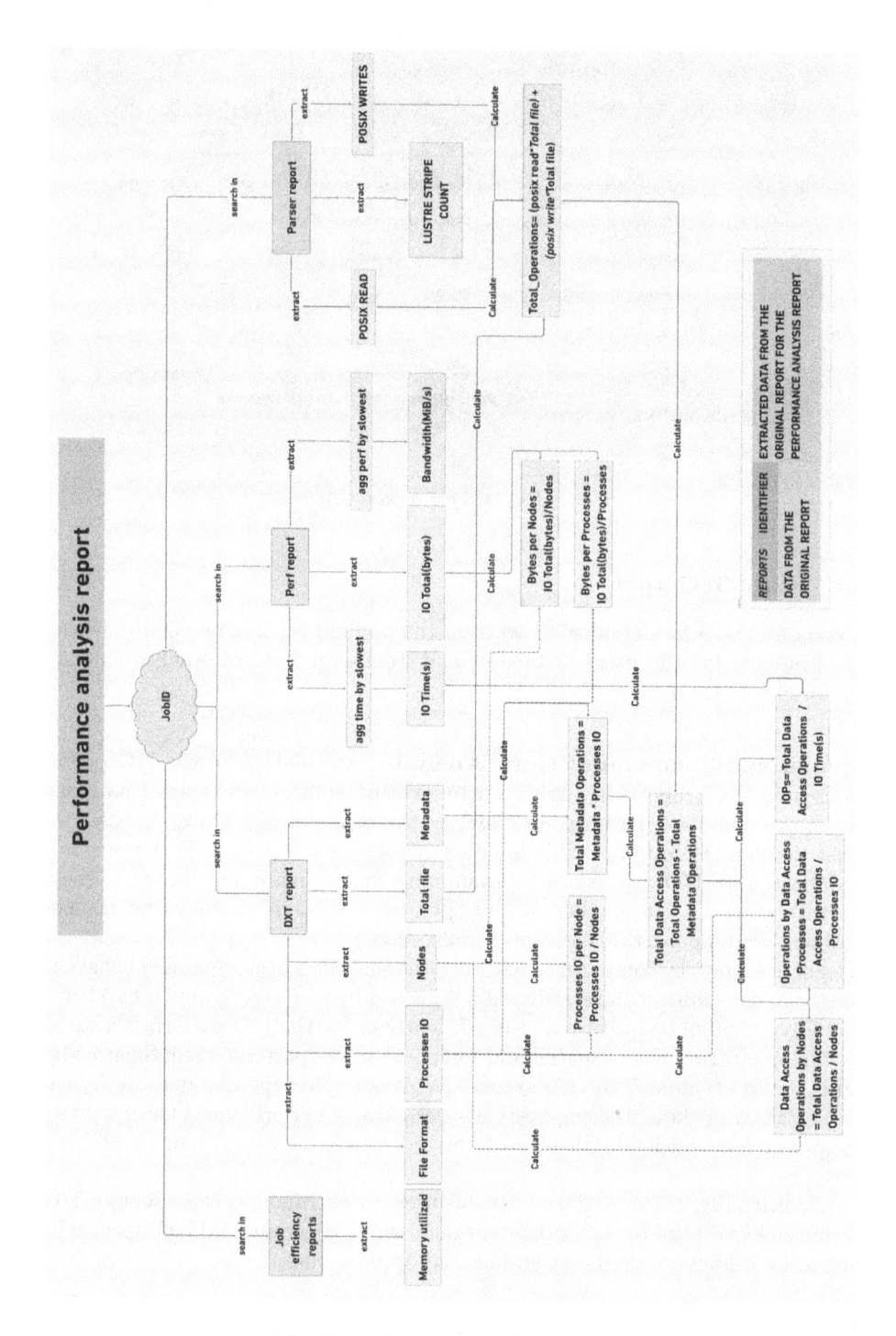

Fig. 5. Performance analysis report

comparing performance metrics to identify optimization opportunities. Figure 5 shows the detailed procedure for monitoring and evaluating job performance in HPC systems. Using data selected from the seff command and the different reports of Darshan tool, we provide a comprehensive view of I/O performance that spans from memory utilization to distribution of I/O operations on the parallel file system. This comprehensive approach allows for pinpointing inefficiencies and understanding the interaction between application I/O patterns and the file system's capabilities.

4 Experimental Validation

In this section, we apply our proposed method in a real DL application. We use Deep Galaxy application that aims to leverage the pattern recognition capability in modern DL to classify the properties of galaxy mergers [14]. The dataset contains 35,784 black and white images from simulations of galaxy mergers of different mass and size ratios. These images are stored in a compressed HDF5 dataset, with an internal structure of 36 folders each with 14 subfolders that represent the different positions of the camera and in each subfolder 71 images are stored with a resolution of 1024*1024 pixels each one, for a total dataset file size of 6.1 GiB. Below we show some graphical reporters designed to analyze DeepGalaxy I/O.

Figures 6 and 7 focus on the spatial and temporal pattern of the I/O operations. In Fig. 6, which is a 3D representation, the X-axis, identified as **"Process IO"**, represents the different processes involved in the I/O operations; the Y-axis displays the **"Temporal Order"**, indicating the sequence in which the I/O operations occurred; and the Z-axis is the **"Offset (GiB)"**, representing the position in the file where the read or write operations take place, expressed in Gibytes (GiB). The legend on the right indicates the size of the read requests (**"Read Request Size (KiB)"**) with a color scale ranging from green, red and blue. In Fig. 7, which is a 2D representation, the X-axis again represents "Process IO," while the Y-axis shows the "Offset (GiB)." The color scale remains the same, indicating the size of the read requests. To understand the underlying patterns and detect potential performance bottlenecks, we proceed to perform a comparative analysis of two graphs representing I/O operations.

4.1 Spatial and Temporal Pattern Analysis

The interaction of I/O operations within a distributed computing environment is very complex. This analysis aims to show these operations' spatial orientation and temporal progression. With this information, we aim to gain insight into the efficiency of data management and the effectiveness of resource utilization in HPC systems.

Figure 6 shows the spatial and temporal pattern of DeeGalaxy when the dataset is read from an NFS file system. However, it is important to note that this pattern is similar on the Lustre file system. This is because the dataset is

86   E. Parraga et al.

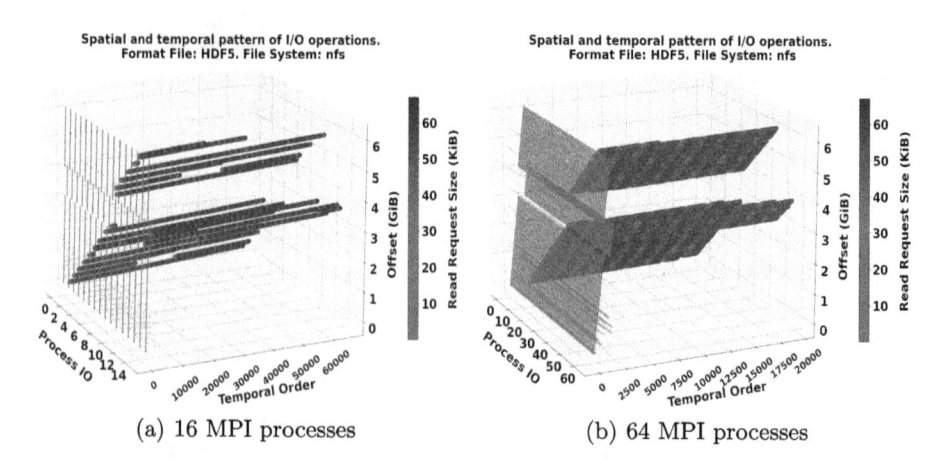

(a) 16 MPI processes

(b) 64 MPI processes

Fig. 6. Temporal I/O pattern on a NFS file system. All processes access a single shared file at a different file offset. 4 MPI processes per compute node.

stored in an HDF5 file; therefore, the observed pattern is the same on both file systems. The only difference will be observed in the timing of the I/O operations, but not in their order.

Figure 6(a) and Fig. 7(a) depicts the variety in the size of I/O requests. The largest request is 66.91 KiB and the smallest request is noticeably smaller, at only 8 bytes. The average size of requests hovered at a median of 6.32 KiB. A total of 1,038,638 read-only I/O operations were performed. These read operations were spread over a set of 4 nodes and orchestrated by 16 I/O processes, following a shared data access pattern. Additionally, we observed a balanced number of I/O operations per process at POSIX-IO level. In terms of data volume, a total of 6.92 GiB is moved on the filesystem, which is 0.82 GiB more than the file size of the dataset.

Figure 6(b) and Fig. 7(b) show variability in the size of I/O requests, with the maximum request size reaching 66.91 KiB and the minimum at only 8 Bytes. The average request size was approximately 5.91 KiB. Throughout the same period, there were 1,226,949 I/O operations performed, this time distributed over 16 nodes and managed by 64 I/O processes. The I/O aggregate data for these operations was also 6.92 GiB. In these pictures, we can also see whether there is an overlap in accesses or if each process read at a different file offset.

Figure 6(a) and Fig. 6(b) depict how requests are distributed across time and file offset, which is critical for understanding access patterns and performance. Increasing from 4 to 16 nodes and from 16 to 64 I/O processes, suggesting that the NFS infrastructure can efficiently handle a growing number of operations without significantly degrading performance.

Figures 7(a) and 7(b) provide a 2D view of the I/O operations per process and their file offset, making it easier to identify whether the I/O processes access different offsets or if there are overlapping accesses to the same offset. In this case, each process reads its part of the file in parallel, so there is no overlap in accesses. This means that if we see any I/O serialization at runtime, it will be due to the underlying I/O system rather than application's I/O pattern.

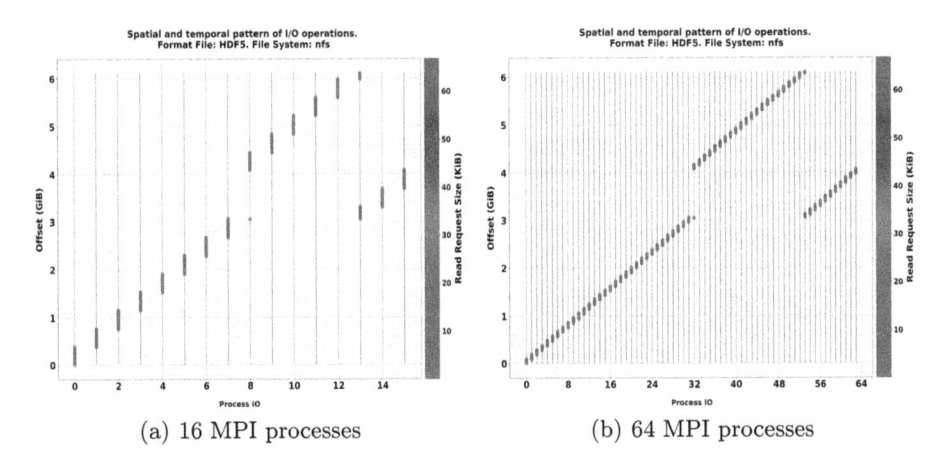

(a) 16 MPI processes

(b) 64 MPI processes

Fig. 7. Spatial I/O pattern on a NFS file system. All processes access a single shared file at a different file offset. 4 MPI processes per compute node.

4.2 DeepGalaxy Performance Analysis

Extending our previous exploration of DeepGalaxy's I/O operations, we present four new plots for performance analysis, each showcasing results for a different number of processes. Furthermore, we executed the application reading dataset from a NFS and a Lustre filesystem. Figure 8(a) and Fig. 8(b) present the I/O performance of the Deep Galaxy application, highlighting the influence of the count of processes and nodes. The x-axis represents the mapping used in each experiment. In all the cases, 4 processes are mapping in each compute node. So the label 16p-4N means a mapping of 16 processes distributed in 4 compute nodes. The y-axis correponts to time in seconds and the secondary y-axis displays the data transfer rate in GiB/s.

Figure 8(a) indicates that NFS filesystem has significant performance fluctuations. At 16p-4N, the average I/O time was 14.26 s with a standard deviation of 4.30. The data transfer rate was relatively stable at 0.48 GiB/s (Std of 0.16). As the process count increased to 32 on 8 nodes, the average I/O time dropped to 8.39 s with a data transfer rate of 0.79 GiB/s. With 48 processes on 12 nodes, the I/O time slightly reduced to 7.54 s, though variability increased (Std of 3.39) with a peak data transfer rate of 1.00 GiB/s. However, at 64 processes on 16 nodes, the I/O time surged to 37.05 s, suggesting instability or potential bottlenecks and the data transfer rate significantly fell to 0.30 GiB/s with Std of 0.22.

In Fig. 8(a), it can be seen that despite initial reductions in I/O time, a significant spike and high variability in I/O time occur for the 64p-16N mapping.

88   E. Parraga et al.

The data transfer rate increases up to 48p-12N, peaking at 1.00 GiB/s, before dropping sharply for 64p-16N. The high standard deviations in both I/O time and data transfer rate at 64p-16N indicate that NFS has problems managing the increasing number of small I/O operations for 64 I/O processes. This is an expected behavior because NFS is not designed to manage parallel I/O accesses.

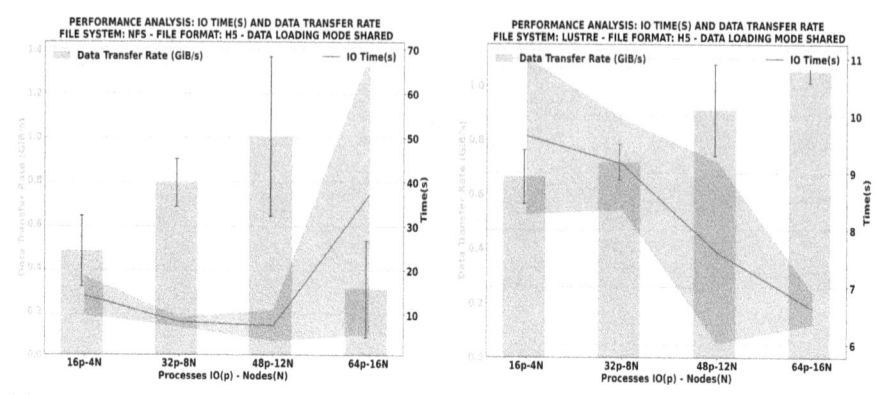

(a) Data Transfer Rate and IO Time. File System: NFS

(b) Data Transfer Rate and IO Time. File System: LUSTRE

Fig. 8. DeepGalaxy I/O Performance by using different number of processes.

Figure 8(b) present the I/O performance of DeepGalaxy when reading dataset from a LUSTRE file system. For the 16p-4N mapping, an average I/O time of 9.64 s is reported with a stable data transfer rate of 0.66 GiB/s. With 32 processes on 8 nodes, I/O time decreases slightly to 9.15 s with an increased data transfer rate of 0.72 GiB/s. For 48 processes on 12 nodes, the I/O time further declines to 7.60 s and the data transfer rate climbs to 0.91 GiB/s. At the highest scale tested, 64 processes on 16 nodes, the average I/O time improves to 6.63 s with a data transfer rate peaking at 1.05 GiB/s.

Therefore, as the number of processes and nodes increases, both filesystems exhibit improved I/O times and similar data transfer rates. However, the variability in I/O times suggests that the Lustre filesystem is more appropriate for a larger number of I/O processes. The comparison clearly favors the Lustre filesystem, which consistently outperforms NFS in all metrics at the scales tested, underscoring its superior management capabilities in handling parallel I/O and load balancing on the data servers.

5 Conclusions

We have presented our method for analyzing I/O traces of DL applications executed in HPC systems. Experimental validation has shown that our method provides useful information to guide users in analyzing the I/O behavior of

DL applications. The temporal and spatial analysis of I/O patterns has offered a comprehensive understanding of I/O behavior, enabling the identification of potential bottlenecks and areas for performance improvement. Additionally, our approach has proven effective in synthesizing and extracting key information from reports generated by monitoring tools like Darshan and the `seff` command from SLURM.

Our research has shown that using a Lustre file system instead of NFS for specific workloads can significantly reduce I/O time and increase data transfer rates. In our experiments, the DeepGalaxy application demonstrated a reduction in I/O time from 37.05 s on NFS to 6.63 s on Lustre when the number of processes was increased from 16 to 64.

Understanding spatial and temporal I/O patterns is fundamental to explaining performance variations. For instance, significant differences in I/O performance were observed for the DeepGalaxy application when using different compute nodes and processes across various file systems. These differences highlight issues such as inefficient data access sequences and uneven distribution of I/O loads. By analyzing these patterns, we can identify and address these issues, leading to improved performance and efficiency.

For future work, we propose to apply the results of our method to address I/O optimization techniques such as caching and prefetching, distributing I/O operations across data servers, and identifying I/O intensity during peak demand periods to avoid I/O system saturation. Although I/O intensity largely depends on the application, it can be managed by implementing adaptive I/O strategies that respond to varying load conditions. For example, during periods of high demand, dynamically reallocating I/O resources and prioritizing critical I/O operations can help mitigate saturation. Additionally, predictive modeling based on historical I/O patterns can forecast high-intensity periods, allowing for preventive adjustments in the I/O infrastructure. By implementing these optimization techniques, we aim to balance the load and improve the overall efficiency of the I/O system.

References

1. Young, T., Hazarika, D., Poria, S., Cambria, E.: Recent trends in deep learning based natural language processing [review article]. IEEE Comput. Intell. Mag. **13**(3), 55–75 (2018)
2. Krizhevsky, A., Sutskever, I., Hinton, G.E.: Imagenet classification with deep convolutional neural networks. Adv. Neural Inf. Process. Syst. **25**, 1097–1105 (2012)
3. Esteva, A., et al.: Dermatologist-level classification of skin cancer with deep neural networks. Nature **542**(7639), 115–118 (2017)
4. Dean, J., Ghemawat, S.: Mapreduce: simplified data processing on large clusters. Commun. ACM **51**(1), 107–113 (2008)
5. Patil, D., Bhatia, P.: High performance computing for big data: methodologies and applications. J. Parallel Distrib. Comput. **103**, 1–19 (2016)
6. Carns, P., et al.: Understanding and improving computational science storage access through continuous characterization. ACM Trans. Storage (TOS) **7**(3), 1–26 (2011). https://doi.org/10.1145/2027066.2027068

7. Lee, J., Bahn, H.: Analyzing data access characteristics of deep learning workloads and implications. In: 2023 3rd International Conference on Electronic Information Engineering and Computer Science (EIECS), pp. 546–551 (2023)
8. Krishnamurthy, R., Heinze, T.S., Haupt, C., Schreiber, A., Meinel, M.: Scientific developers v/s static analysis tools: vision and position paper. In: 2019 IEEE/ACM 12th International Workshop on Cooperative and Human Aspects of Software Engineering (CHASE), pp. 89–90 (2019)
9. Wu, X., et al.: How are deep learning models similar? An empirical study on clone analysis of deep learning software. In: 2020 IEEE/ACM 28th International Conference on Program Comprehension (ICPC), 2020, pp. 172–183 (2020)
10. Bez, J.L., et al.: I/o bottleneck detection and tuning: connecting the dots using interactive log analysis. In: IEEE/ACM Sixth International Parallel Data Systems Workshop (PDSW) 2021, pp. 15–22 (2021)
11. Devarajan, H., Zheng, H., Kougkas, A., Sun, X.-H., Vishwanath, V.: DLIO: a data-centric benchmark for scientific deep learning applications. In: IEEE/ACM 21st International Symposium on Cluster. Cloud and Internet Computing (CCGrid) 2021, pp. 81–91 (2021)
12. Neuwirth, S., Paul, A.K.: Parallel I/O evaluation techniques and emerging HPC workloads: a perspective. In: IEEE International Conference on Cluster Computing (CLUSTER) 2021, pp. 671–679 (2021)
13. Chowdhury, F., et al.: Emulating I/O behavior in scientific workflows on high performance computing systems. In: IEEE/ACM Fifth International Parallel Data Systems Workshop (PDSW) 2020, pp. 34–39 (2020)
14. Cai, M.X., et al.: DeepGalaxy: deducing the properties of galaxy mergers from images using deep neural networks. In: IEEE/ACM Fourth Workshop on Deep Learning on Supercomputers (DLS) 2020, pp. 56–62 (2020)

Smart Cities and E-Government

Industry 5.0. Digital Twins in the Process Industry. A Bibliometric Analysis

Federico Walas Mateo[1]( ) and Armando De Giusti[2]

[1] Universidad Nacional Arturo Jauretche (UNAJ), Buenos Aires, Argentina
`fedewalas@gmail.com`
[2] III-LIDI, Facultad de Informática, Universidad Nacional de La Plata (UNLP), La Plata, Argentina

Abstract. In the context of industrial digitalization, the Industry 5.0 model incorporates digital twins as an innovative tool. This study aims to delve deeper into the concept of digital twins, their integration with the Industrial Internet of Things (IIoT), and how these solutions contribute to bring intelligence into industrial environments.

Digitalization in industry enables connected products and processes to enhance the productivity and efficiency of people, plants, and equipment. The outcomes of these improvements should have widespread impacts on both the economy and the environment. As connected products and processes generate data, this data is increasingly viewed as a fundamental source of competitive advantage, posing new challenges in industrial environments.

The article examines digital twin technology, its integration with IIoT and the intelligence of this devices in the Industry 5.0 or smart manufacturing framework. The focus lies on discussing about the contribution of digital twins for optimizing industrial processes.

The paper reviews relevant articles and conducts a bibliometric analysis of key topics surrounding digital twins as a value-added solution for process optimization within the Industry 5.0 paradigm. The primary findings highlight the growing significance of this subject since 2018, as evidenced by the number of published articles in the Scopus Database. Additionally, the study underscores the complexity of digital twins addressing this issue within the industrial environment.

Keywords: Industry 5.0 · Digital Twin (DT) · Industrial Internet of Things (IIoT) · process industries · Industrial Process Optimization

1 Introduction

Based on the doctoral thesis work [1], it is proposed to move beyond the IIoT paradigm, and analyze architectural alternatives for data integration and the evolution towards artificial intelligence (AI) adoption in the industry. The solution proposed by digital twins (DT) appears as a relevant research point to evolve to more sophisticated the data models, and generate a solution that ease the optimization of the industrial operation [2]. In this framework, there is a strong interest in researching about DT, their integration with the IIoT architecture, and the possibility of working within the framework of prescriptive analytics in order to optimize industrial processes.

DT provide an architecture to integrate data from operations and generate a solution to analyze processes and facilitate their optimization. This work falls within the framework of the ISA 95 standard [3], and RAMI 4.0 (Reference Architecture Model Industrie 4.0), developed by the German Association of Electrical and Electronic Manufacturers (ZVEI) [4].

DT constitute a technological alternative from virtualization and containerization for system management. On the other hand, they facilitate the integration of OT systems with IT in a secure way, generating a technological option to further facilitate the use of IIoT solutions and bring OT to the cloud [5].

In the above article [5], the author points that smart manufacturing systems can connect raw materials, production systems, logistic companies, and maintenance schedules using the capabilities of industrial, IoT. These connections are creating cyber-physical production systems (CPPS) and linking functions across the entire product lifecycle. These connections are possible today because of the advances in digital manufacturing (DM) technologies that can facilitate factory design, redesign, and analysis in CPPS, and help to manage manufacturing process optimization.

On the other hand, it should be considered that the Industry 5.0 (I5.0) model as a concept focuses on three elements, empowering people in the processes, generating more environmentally friendly industrial processes, and facilitating the development of more resilient value chains [6, 7]. It is of interest to analyze the impact of DT to achieve these objectives proposed by the I5.0 model. In this line, [8] points that DT are essential in I5.0 to improve the production cycle. The author highlight that an enormous amount of information gathered from DTs, for instance, can be used to optimize production processes.

This article is linked with a project being developed among the Universidad Autonoma de Barcelona (UAB) from Spain, and research groups from Universidad Nacional Arturo Jauretche (UNAJ), Universidad Nacional de La Plata (UNLP) and Universidad Nacional del Noroeste de la provincial de Buenos Aires (UNNOBA) of Argentina as well as the subsidiary of the Siemens company in Argentina [8]. The focus of this line of research and development is on the use of new technologies, in particular the development of DT applicable in Industry, the agricultural sector, and in Health Management issues.

This paper begins by proposing the conceptual framework, then establish the objectives and hypothesis of the research, then advances to the preliminary findings to go deeper into a bibliometric analysis around the idea of DT as a tool for the optimization of processes within the framework of the I5.0 model in process industries. Methodologically, a technological mapping was carried out through an exercise on Scopus indexed database, whose results was analyzed using bibliometric indicators.

The research was completed through the use of the VOSviewer® 1.6.11 software tool (http://www.vosviewer.com/), to analyze results and make easier to reach the conclusions and possible future lines of research raised from this work.

2 Some Preliminary Findings

A previous bibliographic research was made regarding the issues in which this paper is focused. To start the development of this work, articles with some interesting insights were considered. This evidence seems a good starting point, and then complete the work with bibliometric research.

The first paper considered in this section is [9] in which the authors establish that the concept of DT dates back to 2003, introduced by M. Grieves as a tool to manage the product life cycle. According to this article, digital twins are the backbone of the Industry 4.0 model. The authors describe this solution from three aspects: physical entity, virtual entity, and data connection.

The link between IIoT and DT can be observed in several articles in different ways, [10–14]. These papers give some insights to prepare the bibliometric research about DT in process industries.

[10] Observes that IIoT, DT and progress in mobile networks is currently facilitating the development of decentralized self-managed CPPS within the industry. DT allows mobile networks to provide adaptive and dynamic configurations for cooperative CPPS. Moreover, trustworthy cooperation may be realized with blockchain. The paper presents a solution where blockchain is used to cross-verify and validate newly added blocks with the support of validator nodes.

The work [11] presents a product quality control solution by developing a semanticbased DT information model of terminal device, the flexible adjustment and parameter configuration of terminal device are realized to meet the demands of flexible production and manufacturing. The authors point that the suggested tool to inspect product quality can enhance the utilization of hardware resources and the efficiency of product quality assessment while lowering the overall deployment expenses of the system. Moreover, it offers adaptable compatibility with product variations and various industrial settings.

According to the authors [12], DT technology addresses needs of IIoT by enabling the simulation, monitoring, and optimization of such systems. In this paper, it is analyzed the integration of the formal modeling technique, Petri-nets, within the context of NDTs to model IIoT, approaching data-driven Petri-nets. The results of the research demonstrate the model's effectiveness in executing what-if scenarios based on the network's operational parameters to predict the Packet Delivery Ratio and enable real-time fault detection.

The paper [13, 14] presents an implementation of a digital twin that connects through the IIoT architecture using Node Red.

To advance in the conceptual framework, the work [15] can be cited, where the authors affirm that DT seem to be promising enablers to replicate production systems in real time and analyze them. A DT should be capable to guarantee well-defined services to support various activities such as monitoring, maintenance, management, optimization and safety. Particular focuses of this paper are the degree of integration of the proposed DT with the control of the physical system, in particular with the Manufacturing Execution Systems (MES) when the production system is based on the Automation Pyramid proposed by ISA 95 standard, and the services offered from these environments, comparing them to the reference ones.

3 Objective and Methodology of Bibliometric Research

Having collected some evidence about DT and IIoT in the productive environment, it is aimed to have a deeper understanding of the state of the art of DT and its adoption in process industries, and opportunities to add intelligence to production systems to optimize and add reliability to its operation.

Another observation that is object of this work is what [16] says about studies that have shown that digitization of products and services has become a necessity for a sound industrial ecosystem. However, these requirements and advanced technologies have made the systems more complex and led to many other challenges such as cybersecurity, reliability, integrity, etc. These are the major bottlenecks which needs to be overcome for the successful design and deployment of smart factories.

Then the next step in the research is to develop a bibliometric analysis [17]. Highlights the bibliometric discipline, which is being facilitated by the easy access to articles compiled and indexed in enormous databases, making possible to have data about research facts, like number of authors, keywords, topic, citations, and institutional collaboration, among others.

This paper aims to take in consideration the issues in the above paragraphs, in the framework of the adoption of DT in the industrial environment to add value to the data in process industries environments.

The bibliometric analysis task began with the definition of the search criteria. With the objective established, the Scopus database was searched. This scientific repository was selected on the condition that is one of the broadest scientific data base. On the other hand, taking into account to study a broad scope of articles, "Digital Twin" and "process industry" were selected as the keywords within the search. The search brings all the articles with the keywords included in the title or abstract and/or full text were combined using the Boolean operators "AND".

The query string used for the search was the following:

ALL ("Digital Twin" AND "process industry")

Once the search was concluded, the data found was exported in a CSV file, to analyze results in detail. The export was carried out in two formats, the first with the information related to the complete bibliographic data, and then another file that contains only the keywords and the abstracts of the papers found.

Subsequently, the analysis was completed by incorporating the CSV file from SCOPUS database into the VOSviewer® 1.6.11 software tool to visualize most relevant keywords and its clustering [18].

4 Bibliometric Analysis Results

The search using the methodology described in the previous section was performed on 23rd February 2024, and yielded 93 papers as result. The first parameter to be studied is the evolution of the number of documents per year, which, as shown in Fig. 1. The interest in the subject has been growing since 2018 when the first article about the issue being analyzed appeared. This indicator shows that the subject is a quite novel one with much potential of scholarly research, and growing interest. Something to note is that it is assumed that the decline shown for year 2023 can be associated to articles being in process to be published.

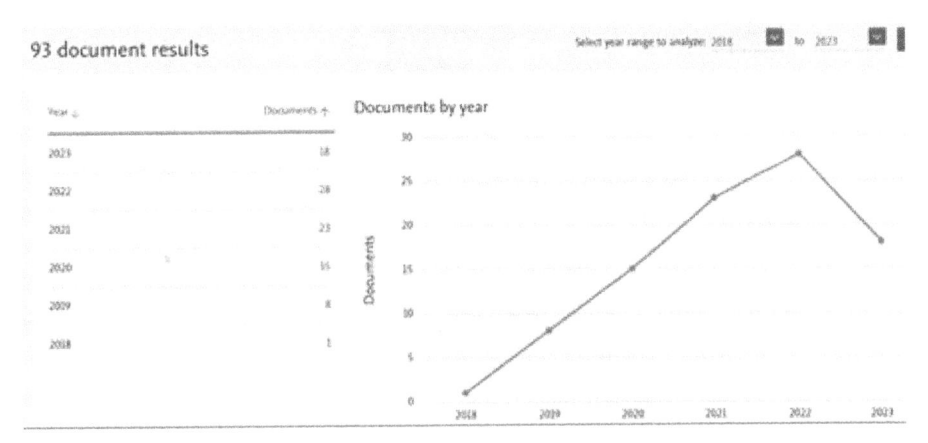

Fig. 1. Evolution in the number of publications about Digital Twins and process industries. Source: Scopus data base.

Then, relevant keywords in the subject were analyzed with the clustering generated by the VOSviewer® 1.6.11 software tool, from the CSV file obtained from Scopus database. Below, Fig. 2 represents the cloud map with relevant words of the articles. This map shows how many times the words appear in the articles and how related are between them using full counting method. It must be noted that in this case the words shown appeared more than 15 times.

The cloud shows seven clusters of words grouped by different colors. At first sight, it could be observed that the term *process industries,* is strongly showed in the map. Among process industries chemical sector seems the most visible in this research, and the reason to be more popular in this kind of production system can be observed associated with the key words *abnormal situation management, complex processes,* and *safety,* which are characteristics that are strongly present in this kind of industry.

Another observation that can be done is associated with adding intelligence to the process, and in this direction the strong presence of keyword *machine learning, agent based,* and *simulation optimization.*

Regarding IIoT and ISA95 standard, although these key words are not visible among top key words, it is possible to relate these issues with key words *data integration* and *data handling* which are visible in the cloud map.

A last insight to be taken into account, from the cloud map, is that linked with Industry 5.0 are the terms *energy transition, operator support,* and *decision making.* The first one is strongly linked to the migration to environment friendly production systems, and the last two refers to empowering the people at the industrial process.

Fig. 2. Cloud map of words in titles and abstracts (full counting), generated with VOSviewer. Source: Scopus Database.

5 Conclusions and Future Research Lines

The first conclusion is the vast amount of information and the potential the subject has in the future. On the other hand, it can see that the concept has a long road to walk to mature and adapt to capture the attention of the industrial world massively.

This research work highlights the novelty of the research topic, validated some premises that motivated this line of work, and give some insights to align the future research steps to learn more about DT adoption in industry and its contribution to add intelligence to process and optimize them. DTs in the process industry reinforces the importance of some technologic topics as IIoT and AI applied to optimize decisions in real time systems.

In line with the above paragraph is the fact that DT could have a relevant role in industries, like oil &gas, mining, energy, among others, where risks for people and environment are very high, more reliable and safer.

The research provided 93 articles to study and go further to collect more knowledge about different issues related to DT and its adoption in the industrial environment, in particular in the process sector. A fast scanning of these papers gave interesting insights for future work.

Future research lines are associated with collaboration with colleagues researchers from UNLP, UAB, and UNNOBA, to study adoption of DT technology in other areas than industry, like agriculture and health services, going further from previous work in the subject.

References

1. Walas Mateo, F.: Tesis Doctoral. Nuevos modelos de negocio en el paradigma Industria 5.0. Inteligencia Artificial y Aprendizaje Automático para optimizar procesos industriales (2023)
2. Shyam Varan, N., Dunkin, A., Chowdhary, N., Patel, N.: Industrial Digital Transformation: Accelerate digital transformation with business optimization, AI, and Industry 4.0. Packt Publishing (2020)
3. American National Standards Institute (ANSI), ISA-95.00.01-2010, ISA-95.00.02-2010, ISA-95.00.03-2013, ISA-95.00.04-2012, ISA-95.00.05-2013, North Carolina USA
4. RAMI4.0. Reference Architecture Model Industrie 4.0. https://www.beuth.de/en/technical-rule/din-spec-91345/250940128. Accessed 1 abril 2023
5. Park, Y., Woo, J., Choi, S.: A cloud-based digital twin manufacturing system based on an interoperable data schema for smart manufacturing. Int. J. Comput. Integr. Manuf. **33**, 1259–1276 (2020)
6. Di Nardo, M., Yu, H.: Special issue. Industry 5.0: the prelude to the sixth industrial revolution. Appl. Syst. Innov. **4**, 45 (2021)
7. Doyle-Kent, M., Kopacek, P.: Industry 5.0: is the manufacturing industry on the cusp of a new revolution? In: Durakbasa, N., Gençyılmaz, M. (eds.) ISPR 2019. LNME, pp. 432–441. Springer, Cham (2020). https://doi.org/10.1007/978-3-030-31343-2_38
8. Mateo, F.W., et al.: Gemelos Digitales: Aplicación en la Industria de Procesos, en la Gestión Sanitaria y en el área Agrícola. Congreso WICC. UNSJB, Puerto Madryn (2024)
9. Alojaiman, B. Technological Modernizations in the industry 5.0 era: a descriptive analysis and future research directions. Processes **11**, 1318 (2023). https://doi.org/10.3390/pr11051318
10. Jiang, Y., Yin, S., Li, K., Luo, H., Kaynak, O.: Industrial applications of digital twins. Phil. Trans. R. Soc. A. **379**, 20200360 (2021). https://doi.org/10.1098/rsta.2020.0360
11. Aloqaily, M., Ridhawi, I.A., Kanhere, S.: Reinforcing industry 4.0 with digital twins and blockchain-assisted federated. IEEE J. Sel. Areas Commun. **41**(11), 3504–3516 (2023). https://doi.org/10.1109/JSAC.2023.3310068
12. Hu, P., He, C., Zhu, Y., et al.: The product quality inspection scheme based on software-defined edge intelligent controller in industrial internet of things. J. Cloud Comput. **12**, 113 (2023). https://doi.org/10.1186/s13677-023-00487-7
13. Kherbache, M., Ahmed, A., Maimour, M., Rondeau, E.: Constructing a network digital twin through formal modeling: tackling the virtual–real mapping challenge in IIoT networks. Internet of Things **24**, 101000 (2023). ISSN 2542-6605. https://doi.org/10.1016/j.iot.2023.101000
14. Cimino, C., Negri, E., Fumagalli, L.: Review of digital twin applications in manufacturing. Comput. Ind. **113**, 103130 (2019). https://doi.org/10.1016/j.compind.2019.103130. ISSN 0166-3615
15. Ngonidzashe, M.B., Tuncay, E.: A simple node-RED implementation for digital twins in the area of manufacturing. Trends Comput. Sci. Inf. Technol. **8**(2), 050-054 (2023). https://doi.org/10.17352/tcsit.000068
16. Muhuri, P. K., Shukla, A.K., Abraham, A.: Industry 4.0. A bibliometric analysis and detailed overview. Eng. Appl. Artif. Intell. **78**, 218–235 (2019)
17. Merediz-Solà, I., Bariviera, A.F.: A bibliometric analysis of bitcoin scientific production. Res. Int. Bus. Financ. **50**, 294–305 (2019)
18. Van Eck, J., Waltman, L.: Manual for VOSviewer version 1.6.15. Universiteit Leinden (2020)

Visualization

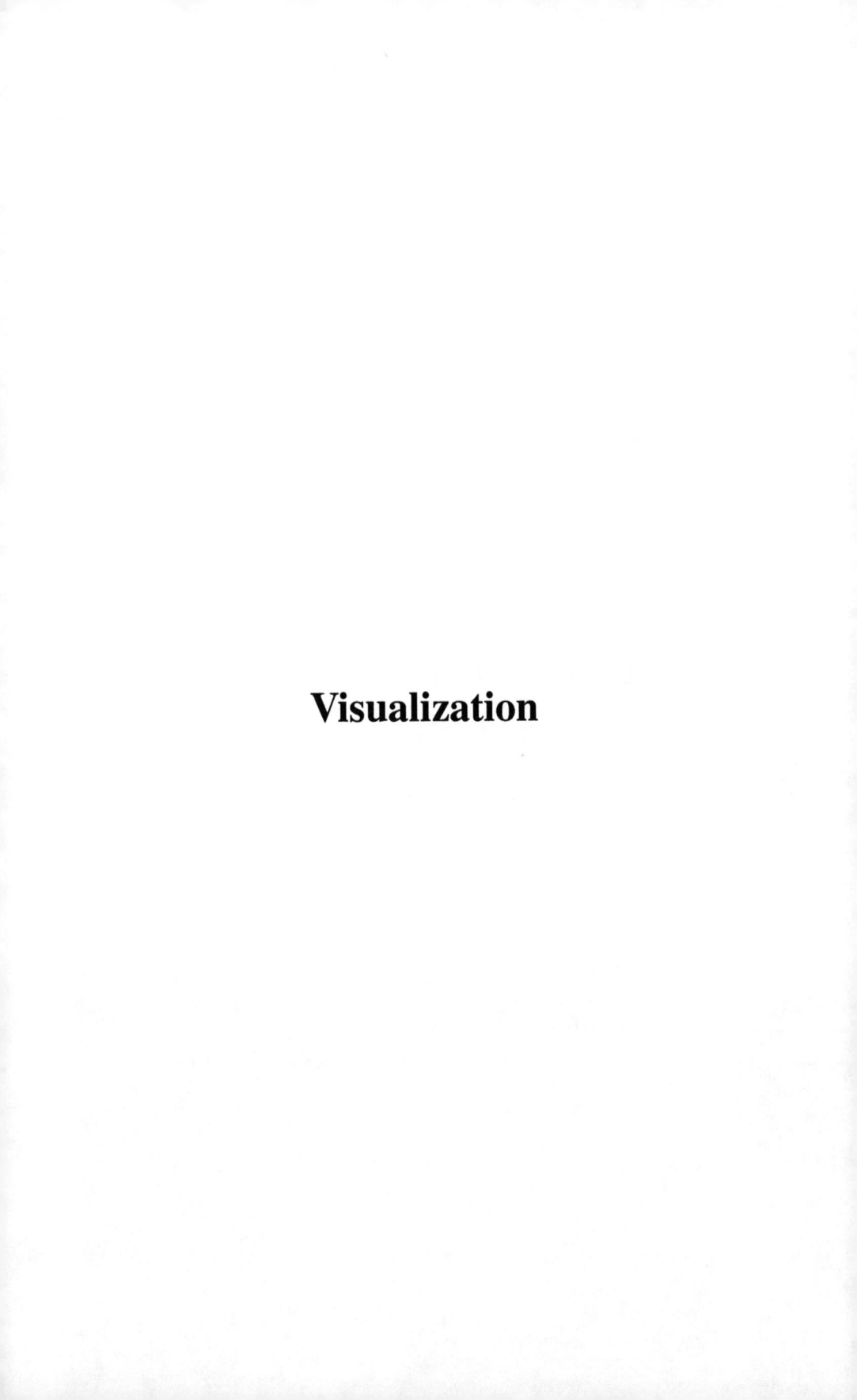

An ABMS COVID-19 Propagation Model for Hospital Emergency Departments

Morteza Ansari Dogaheh[1(✉)], Manel Taboada[2], Francisco Epelde[3], Emilio Luque[1], Dolores Rexachs[1], and Alvaro Wong[1]

[1] Computer Architecture and Operating Systems Department, University Autonoma of Barcelona, Bellaterra, Barcelona, Spain
morteza.ansaridogaheh@autonoma.cat, emilio.luque@uab.es,
{Dolores.Rexachs,Alvaro.Wong}@uab.cat

[2] Escuelas Universitarias Gimbernat: Computer Science School, University Autonoma of Barcelona, Barcelona, Spain
manel.taboada@eug.es

[3] Medical Department, Hospital Universitari Parc Tauli, Sabadell, Barcelona, Spain
fepelde@tauli.cat

Abstract. The spread of COVID-19 between different agents in a hospital emergency department can be simulated by modeling the interactions between the agents and the environment. In this research, we use Agent Based Modeling and Simulation techniques to build a model of COVID-19 propagation based on an Emergency Department Simulator which has been tested and validated previously. The benefits of ABM include its ability to simulate complex systems, its flexibility, and its ability to model the interactions between different agents in the system. The obtained model will allow us to build a propagation simulator that enables us to build virtual environments with the aim of analyzing how the interactions between agents influence the rate of virus transmission. The model can be used to study the effectiveness of different interventions, such as social distancing, wearing masks, and vaccination, in reducing the spread of COVID-19.

Keywords: Emergency Department Simulator · Agent Based Modeling and Simulation · COVID-19 Propagation

1 Introduction

COVID-19, also known as coronavirus disease 2019, is an infectious disease caused by the SARS-CoV-2 virus. The virus spreads mainly through respiratory droplets produced when an infected person coughs or sneezes. Symptoms of COVID-19 can appear 2 to 14 days after exposure, can range from mild to severe, and can include fever, cough, shortness of breath, fatigue, muscle aches, headache, sore throat, congestion or runny nose, nausea or vomiting, and loss of taste or smell. Most people who are infected with COVID-19 will recover without requiring hospitalization. However, some people, especially older adults and those with underlying medical conditions, may develop severe illness and require hospitalization. In some cases, COVID-19 can lead to more serious complications, such as pneumonia, acute respiratory distress syndrome, and death. The best way to protect yourself from COVID-19 is to get vaccinated and boosted. You should also wear a mask in public indoor settings, wash your hands frequently, and avoid close contact with people who are sick.

Emergency Departments (ED) are one of the most complex and dynamic healthcare systems, receiving an increasing demand, and usually being overcrowded. ED has a continuous activity, operating 24 h a day, every day of the year. The existence of a communicable disease within the emergency services, especially a seasonal contagious disease [1], can have a negative influence on the service offered, and increases the risk of both, the infection of health personnel and of worsening of the underlying disease of the patients attended in the ED. The resource planning of ED is complex because its activity is not linear, and it varies depending on time, day of the week, and season.

The COVID-19 pandemic has had a significant impact on emergency departments around the world. Some of the most common problems generated by the virus in EDs include the significant increase of patient arrival, the increasing risk of exposition to COVID-19, the mandatory changes in the clinical practice for mitigating the spread of the virus, and the increase of financial pressure, due to the necessity of increase of both, the number of staff and the purchase of PPE (personal protective equipment). In addition to these problems, the COVID-19 pandemic has also had a psychological impact on ED staff. Many staff members have reported feeling stressed, anxious, and exhausted.

Agent-Based Modeling and Simulation (ABMS) is a good solution to study the propagation of COVID-19 in EDs because it allows a detailed and realistic representation of the complex interactions between patients, staff, and the environment. This is important because the spread of COVID-19 is not simply a matter of individual risk factors, but also of the way that people interact with each other in the ED setting. ABMS models can be used to simulate the spread of COVID-19 in a variety of ways, including the behavior of patients, the environmental factors (as ventilation and air circulation) and tracking the movement of patients and staff through the ED.

Using ABMS, we can gain a better understanding of how COVID-19 spreads in EDs and let to identify ways to mitigate the spread of the virus. This information can then be used to improve infection control measures and to make the ED a safer environment for patients and staff. The model can be used to explore different scenarios and evaluate the effectiveness of different public health strategies.

In this paper, we begin by describing the previous model of the simulator, along with related works by other researchers. We then propose the essential elements and variables that have an impact on our model. Our aim is to demonstrate the state transition table and present the computational model and implementation of the COVID-19 touch transmission model, along with the results of the experiments based on it. We also put forward the conceptual modeling of the COVID-19 aerial transmission model.

2 Related and Previous Works

We based our model on a previous Emergency Department model (ED-Simulator) that was developed as part of previous research work to simulate an ED according to Spanish standard emergency department [2–4] by the research group High Performance Computing for Efficient Applications and Simulation (HPC4EAS), of the Universitat Autonoma of Barcelona (UAB).

We aim to continue their work by improving and tuning the T-simulator which is developed for MRSA (methicillin-resistant Staphylococcus aureus) study to measure the spread of COVID-19 in an ED. Jaramillo et al. [5] based their contribution on simulating MRSA transmission among active and passive agents in a hospital emergency department. They have performed the model on an existing ED-simulator which has been produced earlier in our research group by other researchers. They present a mathematical model that can be used to predict the spread of nosocomial infections (NIs) in hospital emergency departments. The model considers the following factors: a) the number of patients and staff in the ED; b) the frequency of contact between patients and staff; c) the infectiousness of the NI agent; d) and the effectiveness of infection control measures.

The model was used to simulate the spread of MRSA in a hypothetical ED. The proposed simulator was named T-simulator. The results showed that the number of MRSA infections was highest when the number of patients and staff was high, the frequency of contact was high, and the infectiousness of MRSA was high. The effectiveness of infection control measures was also found to have a significant impact on the spread of MRSA.

The model is a useful tool for understanding the factors that contribute to the spread of MRSA in EDs. It can also be used to assess the impact of new infection control measures on the spread of MRSA. Some features of the model are:

- It is a dynamic model, which means that it can be used to simulate the spread of MRSA over time.
- It is a stochastic model, which means that it takes into account random events, such as the chance of contact between agents.
- It is a customizable model, which means that it can be used to simulate different scenarios, such as different numbers of patients and staff, different frequencies of contact, and different levels of infection control.

The model was used to study a number of different aspects of the spread of MRSA in EDs, including the impact of different infection control measures on the spread of MRSA, the role of the environment in the spread of MRSA, and the impact of patient demographics on the spread of MRSA. The model was a valuable tool for understanding the spread of MRSA in EDs and for developing effective infection control measures.

Wang et al. [6] carry out a study using ABM to simulate the spread of COVID-19 in an academic hospital emergency department and identified best practices to reduce transmission of the virus. Using an agent-based model that simulates patient movements and interactions, the study finds that exceeding the service capacity of 13 patients per hour leads to dramatic increases in contact time, exposure dose, and infection risk growth rates. The model also shows that waiting areas and high-traffic corridors contribute to 66.5% of total exposure dose due to crowding effects. These results highlight the need for careful capacity planning and layout optimizations in fever clinic design to mitigate nosocomial infection risks during an infectious disease outbreak. The paper provides a data-driven approach to evaluate clinic operational decisions and spatial designs with respect to measurable public health impacts.

Hinch et al. [7] used ABM to demonstrate how the virus is transmitted and how different intervention strategies can evaluate the transmission. An agent-based model named OpenABM-Covid19 for simulating the spread of COVID-19 and assessing different interventions like lockdowns, testing, contact tracing, and vaccinations. The model was applied to the first wave of the COVID-19 epidemic in England and with minimal calibration provided a close fit to observed data on deaths, hospitalizations, and seroprevalence. It demonstrates the model's ability to accurately simulate COVID-19 transmission dynamics and evaluate complex intervention packages, making it a useful tool for policymakers weighing options to control epidemics. However, it does not include some real-world complexities, the population structure is simplified and does not capture meta-population dynamics or indoor transmissions like EDs.

Ponsford et al. [8] discuss strategies implemented in an emergency department to prevent transmission of COVID-19 during the pandemic. The most important strategy was developing an efficient triage screening process to quickly identify and isolate potential COVID-19 patients. This involved screening all patients with questions about symptoms and exposure risk upon arrival, then directing them to designated treatment areas based on COVID-19 status. Data showed this triage process successfully captured over 90% of patients who screened positive and later tested positive for COVID-19. The screening and separating of patients, along with proper PPE use, training, and environmental changes helped minimize COVID-19 transmission to healthcare workers (HCW), staff, and other patients in the ED. The screening processes, patient/staff cohorting, and measured transmission rates can help make the ABMS model more closely mirror demonstrated real-world practices and effects during COVID-19.

In the work carried out by Howick, et al. [9] various infection control strategies are evaluated through agent-based modeling in a care home setting, routine testing of staff for COVID-19, in conjunction with measures like hand hygiene and use of PPE, appears to be the most effective and practical approach for controlling spread of the disease. The model predicts that weekly testing of staff, along with isolating symptomatic residents and restricting visitors, substantially reduces infections compared to just implementing those measures alone. Routine testing of residents did not provide additional benefit. The results highlight the importance of reducing transmission risk per contact via interventions like handwashing and PPE for controlling spread in care homes. Agent-based model of COVID-19 spread in a care home setting could help improve an ABM for emergency departments by demonstrating how to capture individual heterogeneity, model distinct staff and patient agents, incorporate pre-symptomatic transmission, evaluate various interventions like PPE and testing, perform uncertainty analysis on key parameters like transmission risk, account for stochasticity through multiple simulations, and validate against real outbreak data. Applying these modeling techniques for distinct features of the ED, like waiting areas and patient/staff flows, could lead to a more useful stochastic model to assess interventions for controlling COVID spread in emergency departments.

3 New Medical Research Contributions and the "Covid 19 Airborne Transmission"

There are a lot of challenges in studying the virus behavior in different situations. Initially, based on the available evidence and the analogy with other respiratory viruses, the World Health Organization (WHO) and other health authorities recommended the "Touch transmission model", which assumes that the virus is mainly transmitted through direct contact with infected people. According to this model, the main preventive measures are hand hygiene, physical distancing, and wearing masks when close contact is unavoidable. However, as the pandemic progressed and more data became available, some researchers started to question the adequacy of the touch transmission model and proposed an alternative hypothesis: the "Airborne transmission model". This model suggests that the virus can also be transmitted through inhalation of small respiratory droplets or aerosols that can remain suspended in the air for longer periods and travel over greater distances than the larger droplets that fall to the ground within 1–2 m. According to this model, the main preventive measures are improving ventilation, avoiding crowded and poorly ventilated spaces, and wearing masks in all indoor settings. The airborne transmission model has gained increasing support from the scientific community in recent months, as several studies have shown the presence and viability of the virus in aerosols under various conditions, and the occurrence of outbreaks in scenarios that are consistent with airborne transmission, such as choir rehearsals, restaurants, and health centers like emergency departments. Moreover, some modeling studies have estimated that a significant proportion of infections can be attributed to aerosol transmission, and that reducing aerosol exposure can have a substantial impact on the pandemic control [10].

In July 2020, a group of 239 experts from 32 countries published an open letter in the journal Clinical Infectious Diseases, urging the WHO and other health authorities to acknowledge the potential role of airborne transmission and revise their guidelines accordingly [11]. The WHO responded by acknowledging that airborne transmission cannot be ruled out, especially in certain settings, but maintained that more research is needed to confirm its extent and implications. The debate between the proponents of the touch and airborne transmission models is still ongoing, as new evidence emerges and new variants of the virus spread. However, most experts agree that a comprehensive approach that considers both modes of transmission and implements multiple layers of protection is the best way to prevent virus transmission.

Later on, and in 2023 researchers in [12] represented a hybrid agent-based model to study the airborne spread of COVID-19. The model considers virus transmission through diffusion in the air and incorporates various factors such as movement, infection susceptibility, recovery rate, vaccination, and the impact of new variants. The model is benchmarked against the infection data of the UK, Italy, and France, showing good practice. The paper explores the effects of parameters such as movement, infection rates, incubation period, duration of infection, and vaccination on the infection spread. It also emphasizes the importance of wearing masks in reducing the spread of the virus and highlights the role of aerial transmission and movement of individuals in the spread of the virus. The study suggests that a shorter incubation period and early detection of infection are key factors in limiting the spread, and it emphasizes the importance of implementing measures to limit infection spread among younger age groups.

4 Emergency Department Process and Model with COVID-19 Propagation

4.1 Basis of ABMS

In ABMS the behavior of agents can be modeled using a **state transition table**, which is a table that shows the possible states of an agent and the conditions or events that trigger a change from one state to another. A state is a condition or situation that an agent can be in at a given time, such as susceptible, exposed, infected, etc. A transition is a change from one state to another that occurs when an input or event triggers it, such as a timer, a message, a collision, etc. A possible state transition table is (Table 1):

Table 1. A state transition table

Current State	Input	Next State	Probability
S_x	I_0	S_x	P_3
S_x	I_0	S_y	P_1
S_x	I_0	S_z	P_2

The content of the table means that, an agent in state S_x receiving input I_0 may move to either state S_y, state S_z, or remain in the same state, with a probability of P_1, P_2, and P_3 respectively. The transitions also can be presented graphically through a **state transitions diagram,** what consists in a combination of nodes and edges, where each node represents a state, and each edge represents a transition. Figure 1 represents a state transition diagram for the previous table.

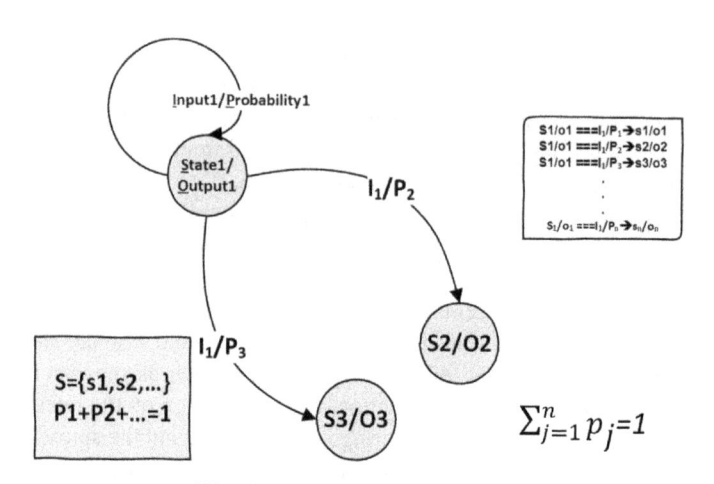

Fig. 1. A state transition diagram

4.2 The COVID-19 Touch Transmission Model

The propagation model of COVID-19 in ED is a playground of different elements like agents, environments, behaviors, and rules. The model includes different types of agents (patients and healthcare workers, like doctors, nurses, and admission staff), the state variables (patient's acuity level: from 1 to 5; current place in ED: Admission, Waiting room1, Triage, Waiting room2, Treatment Area; fragility: from 1 to 3, a measure that quantifies how vulnerable a person is to the virus, based on their age, acuity level and underlying diseases; testing status: received vs. did not receive COVID-19 test; test result: positive, negative; infection status: susceptible, exposed, infected.), and different inputs or events that trigger state transitions (such as contact with other agents, testing positive for COVID-19, receiving treatment, etc.). We may also need to consider the effects of interventions such as social distancing, wearing masks, and other ICMs (infection control measures) on the probabilities of state transitions.

The model also includes environment (ED and its subareas), behaviors (actions taken by agents), interactions (contacts between agents), rules (conditions that govern how agents change their attributes or behaviors), parameters (values that quantify how likely or how often certain rules apply), outputs and outcomes (variables that measure what happens during and at the end of a simulation run), and calibration and validation (processes to ensure the accuracy of the model).

Table 2 presents an example of a state transition table for patients, where q is the probability of infection given contact with an infected agent; r is the probability of testing positive given exposure; s is the probability of recovery given treatment. These probabilities may depend on various factors such as age, predisposition, fragility, comorbidities, viral load etc. The state transition table is a way of specifying a finite-state machine (FSM), which can be either a Moore machine or a Mealy machine. In a Moore machine, the output depends only on the current state of the machine, while in a Mealy machine, the output also depends on the inputs to the machine. In our case, we are using a Mealy machine because the outputs (such as the number of new infections) depend on the inputs (such as the transmission probability).

Table 2. State transition table for a HCW

Current state	Input/Event	Next state	Probability
Susceptible	*Contact with infected agent (patient/hcw)*	*Exposed*	q
Susceptible	*Contact with susceptible agent (patient/hcw)*	*Susceptible*	1-q
Exposed	*Test positive for COVID-19*	*Infected*	r
Exposed	*Test negative for COVID-19*	*Susceptible*	1-r
Infected	*Receive treatment*	*Recovered*	s
Infected	*Do not receive treatment*	*Infected*	
Healthcare worker protected with PPE	*Contact*	*Healthcare worker protected with PPE*	1

110   M. Ansari Dogaheh et al.

As shown in Fig. 2, a patient with no infection having contact with others would change their state based on receiving the virus or not. The patient will repeat their status S while there is no contact or a contact with a susceptible patient. If a contact with an infected occurs the possibility of acquiring the virus should be calculated. The state is changed to E, and a covid19 test is given. If the result is positive the state will change to I, otherwise, we repeat the state S. While the state of the patient is I, they receive treatment to remediate their Corona. After receiving the treatment, the state of the patient will be R or X whether they are recovered or expired. Once the patient is recovered, they may be discharged, or they may stay at the hospital to continue to get treatment for other illnesses.

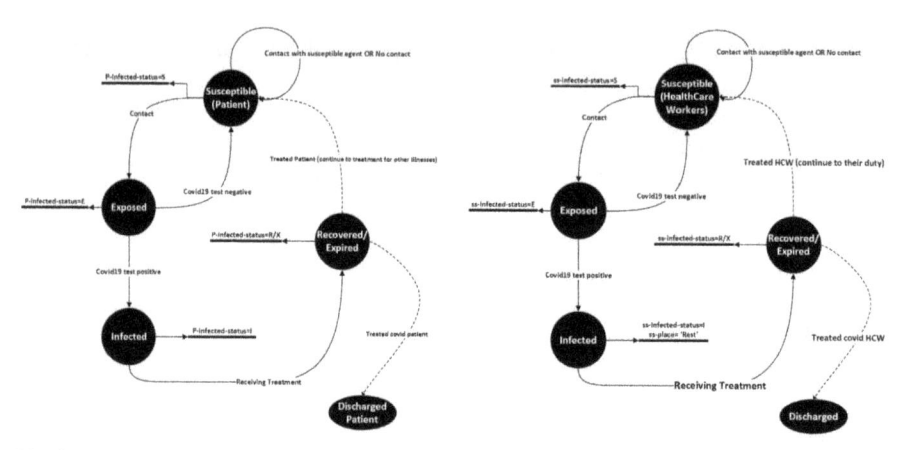

Fig. 2. Virus transmission evaluation for a patient

Fig. 3. Virus transmission evaluation for a HCW

This model of the emergency department examines the dynamics between various active agents in healthcare delivery (Fig. 3): the admission staff, triage nurses, patients, physicians, nursing staff, lab technicians, cleaning staff, and auxiliary workers. Each participant plays a distinct role in the process of providing care. This study assesses three preventative measures:

1. Handwashing, deemed the most crucial, mandates healthcare workers to cleanse their hands prior to and following any patient's physical touch.
2. Hand sanitization using an alcohol-based solution is also essential before and after patient contact. While not a replacement for handwashing, its effectiveness is enhanced when used in conjunction.
3. The use of isolation materials: this is required only when they attend an isolated patient.

These protocols are critical in maintaining hygiene standards and preventing the spread of infections within healthcare settings.

5 Simulation of the "Covid 19 Touch Transmission Model"

5.1 Health Care Procedure

Upon entering the emergency department, patients first visit the admission area. The admission staff request their health card and log their arrival. Subsequently, the patient is directed to the waiting area, known as WR1 in our model, to await the triage process. When a triage nurse becomes available, they call the patient, measure their vital signs, and gather additional information to determine the patient's acuity level. The Spanish ED uses an acuity scale with five levels, ranging from I (highest) to V (lowest). Patients with an acuity level of IV or V wait in a separate area named WR2 for diagnosis and treatment, while those with levels I, II, or III are immediately assigned to a carebox where all phases of diagnosis and treatment occur, except certain specific tests. The diagnosis and treatment process consists of four stages: 1) Initial evaluation; 2) Laboratory testing; 3) Application of medication or treatment; 4) discharge from the ED (to home or hospital). When a doctor is available, they consult with the patient to decide the next course of action. Laboratory tests and treatments can be repeated as necessary.

Table 3. Possible interactions

Possible interaction between agents
Interaction between a Contaminated Patient (active agent) and a ss(HCW)
Interaction between a Contaminated Patient and a susceptible patient
Interaction between a Contaminated Patient and a carebox (passive agent)

Table 4. Possible ways of transmission among agents

Transmission agents	Location	Type of transmission
Patient to patient	Waiting room1, wr2	Direct transmission
Patient to hcw	Admission, triage, carebox	Direct transmission
Patient to carebox	Carebox	Indirect transmission
Careboxb to patient	Carebox	Indirect transmission

Once the treatment is complete and no further lab tests are required, the doctor will recommend the patient to be discharged from the ED. For patients with acuity levels IV and V, their interactions with the doctor will take place in dedicated attention boxes, and they will stay in a designated waiting area during periods of no interaction (between each phase). Our focus is on the stages where agents, meaning patients and healthcare workers, have physical contact. The propagation of Covid19 can occur in several ways (as outlined in Tables 3 and 4). If a patient is a potential transmitter and comes into contact with a susceptible individual, there's a likelihood of transmission, leading to a change in the susceptible agent's state. The variable 'p-real-infected' is used to track changes in the agents' states. To determine if an agent contracts the coronavirus, we use a probability distribution that considers the likelihood of the transmission vector passing on the microorganism and the probability of a susceptible patient acquiring it.

5.2 Implementation of Propagation Model

Healthcare settings are a major risk factor for COVID-19 transmission, especially during close contact between patients and staff. While every interaction carries some risk, the actual outcome depends on several factors.

- Patient susceptibility: This includes their overall health, any underlying conditions, and whether they already have COVID-19.
- Staff protection: Measures like masks, PPE, hand hygiene, and proper disinfection procedures all play a crucial role in reducing transmission risk.

To accurately assess the likelihood of someone getting infected, we need a model that considers these factors in each interaction. This model would look at the patient's vulnerability and the effectiveness of the staff's preventive measures.

The model considers the parts of the comprehensive care process where agents interact, as these moments can facilitate contact propagation. The process initiates when a patient enters the ED and proceeds to the admission area. The admission staff request their health card and record their arrival. If the patient has been admitted previously, the ED has access to their clinical history and can identify their infectious status. Following this, the patient moves to the waiting area to await the triage process. When a triage nurse is free, they contact the patient via the Information System (IS), take their vital signs (an interaction), and gather additional information to determine the patient's acuity level, infectious status, and other crucial data. Patients with an acuity level of IV or V wait for their diagnosis and treatment in a waiting room WR2 (there is also another waiting area in zone B that we call it "Area-B-wr" in our simulator), while those with levels I, II, or III are assigned to a carebox (zone A) where all phases of diagnosis and treatment occur, except for certain specific tests. Similarly, when a patient's infectious state is identified as infected, the IS records this status and triggers an alert in the patient's clinical history.

5.3 Touch Propagation Simulator

The likelihood of contracting the virus depends on factors such as the location, agent's infectious status, and characteristics of the agent, including its fragility and predisposition, as well as preventive measures like wearing a mask. When a patient is located in an area (a carebox/a waiting room etc.) or interacts with others, we evaluate if the patient is susceptible (var p-infect = 0) and agents/carebox/ is contaminated. With some probability based on predisposition, fragility, ICM, etc., the patient can become infected. Then the patient's p-infect variable is updated. Fragility (1–3) is a measure that quantifies how vulnerable a person is to the virus, based on their age, acuity level and underlying diseases. We propose a function that calculates the fragility score of a person using these variables. Predisposition (Yes/No) is also a term that describes the likelihood of a person contracting the virus based on certain factors, such as having a compromised immune system or a chronic condition.

The model also outlines the conditions that need to be satisfied for the virus to develop in a patient's body and for it to be transferred to others. The model further describes the different stages of patient care, from admission to treatment and discharge according to our previous simulator, as well as the potential outcomes of infection (recovery or death).

Implementing a model in the NetLogo platform involves a series of structured steps to ensure that the simulation runs effectively. Initially, one must define the purpose of the model and the phenomena it aims to simulate. Following this, the creation of agents and the environment within NetLogo is crucial, as they are the primary components of any agent-based model. The agents' behaviors, interactions, and the rules governing the environment must be programmed using NetLogo's coding language. It's essential to utilize the interface elements such as buttons, sliders, and monitors to control the simulation and visualize the results. Testing the model with various parameters and refining it based on the outcomes is a continuous process. Here are multiple steps to deploy the model on the Netlogo platform, similar to the approach taken with our earlier ED-simulator:

1) Initialization: The model creates a population of agents with attributes such as age, sex, location/place, health status (susceptible, infectious, recovered), viral load (the percentage of incoming infected patients), diagnosis (true or false), and intervention status (tested, isolated, wearing masks, PPE, vaccinated, etc.). The model also initializes the parameters for transmission probability (based on viral load and research results), predisposition, fragility, disease progression (based on age), intervention efficacy (based on literature), and intervention coverage (based on user input).
2) Simulation: The model simulates the daily interactions-movements of agents in different subareas of our proposed ED. The model updates the health status of each agent based on their exposure to infectious area and their disease progression.
3) Analysis: The model analyzes the simulation results to estimate the epidemic trends and outcomes under different scenarios. The model can also compare the results with real data to validate the model performance and calibrate the model parameters.

Certain initial parameters for the model are entered directly through text files, while others, such as the percentage of incoming infected patients, predisposition percentage, percentage of fragile patients, HCW hand wash probability and its effectiveness, hand disinfection and its effectiveness, isolation material percentage, and its effectiveness can be adjusted in the graphical interface. It's feasible to assign values to each of the three preventive policy actions: hand washing, hand disinfection, or the use of isolated material, and to assign an effectiveness value to each action.

5.4 Experiments with the ED Simulator

The ED Simulator, incorporating a Covid-19 touch transmission model, provides a compelling tool for understanding the spread of the virus through contact. These simulations are crucial for policymakers and health professionals, as they can inform decisions on public health measures and illustrate the potential outcomes of various strategies. The results from such simulations underscore the significance of proactive measures and provide a framework for predicting the effects of interventions on infection rates and overall case numbers.

The simulator begins with preset values for certain variables, such as the percentage of infected patients arriving at the ED and the period of monitoring has been selected for 24 days. The simulation does not consider patient interactions with the admission staff, as the physical contact is minimal compared to interactions with other healthcare staff. For a detailed analysis of person-to-person Covid19 transmission, we will assume in these runs that the only possible transmission routes are direct transmission between patients and other patients, triage nurses, doctors, and auxiliary staff, as well as indirect transmission from patients to careboxes, and subsequently from careboxes to a patient or a healthcare worker.

According to a simulation scenario which is represented in Table 5, the emergency department would receive 1534 incoming patients during a certain period, out of which 5% (77 patients) are infected with COVID-19. The simulation then predicts that the virus transmission at the ED would result in 124 new infections among the patients and staff, adding to the burden of the system and the risk of complications.

Table 5. Different simulation scenario

Simulation Scenario	Incoming Patients	COVID-19 Patients at arrival	New Infections
5% No protection	1534	5% (77 patients)	8.5% (124 p)
10% No protection	1534	10%	10.18%
20% No Protection	1534	20%	14.5%
40% No protection	1534	40%	19.24%

If the scenario were repeated with 20% of covid19 patients coming in, the increase in new infections would be 10.5%. (Table 5)

An ABMS COVID-19 Propagation Model for Hospital Emergency Departments 115

'Incoming Patients', 'COVID-19 Patients at arrival', 'New Infections' by 'Simulation Scenario'

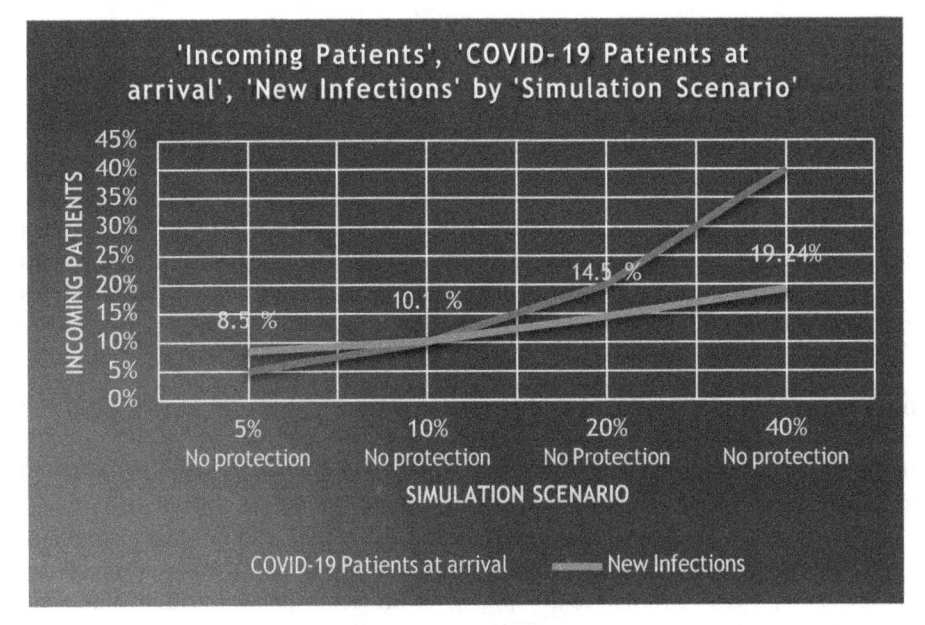

Fig. 4. Simulation scenario graph

The graph shown in Fig. 4 illustrates the relationship between the percentage of COVID-19 patients at arrival and the corresponding new infections at the emergency department (ED) in different simulation scenarios.

- The x-axis represents the percentage of COVID-19 patients at arrival, ranging from 0% to 40%.
- The y-axis represents the percentage of new infections among patients and staff at the ED, ranging from 0% to 20%.

The graph (Fig. 4) shows four simulation scenarios:

1. Scenario: 5% No protection

   - Incoming Patients: 1534
   - COVID-19 Patients at arrival: 5% (77 patients)
   - New Infections: 8.5% (124 patients)

2. Scenario: 10% No protection

   - Incoming Patients: 1534
   - COVID-19 Patients at arrival: 10%
   - New Infections: 10.18%

3. Scenario: 20% No Protection

   - Incoming Patients: 1534
   - COVID-19 Patients at arrival: 20%
   - New Infections: 14.5%

4. Scenario: 40% No protection
    - Incoming Patients: 1534
    - COVID-19 Patients at arrival: 40%
    - New Infections: 19.24%

From the graph (Fig. 4), we can observe that as the percentage of COVID-19 patients at arrival increases, the percentage of new infections also increases. This indicates a direct correlation between the two variables. Additionally, the rate of increase in new infections appears to be non-linear, with a steeper increase as the percentage of COVID-19 patients at arrival increases from 0% to 20%.

6 Future Works and Contributions

In this paper, we proposed an agents-based model of aerial propagation of COVID-19 in a hospital emergency department. The model has been designed based on an Emergency Department Model and Simulation and a MRSA propagation Model and Simulation, both developed in previous research by the Universitat Autonoma of Barcelona research group High Performance Computing for Efficient Applications and Simulation (HPC4EAS), with the participation of ED staff of the Hospital of Sabadell (Spain). This has the advantage of having been verified and validated in several cycles or iterations, considering a wide variety of data and configurations. After a careful analysis of the care process the model obtained in the previous works has been upgraded identifying the types of transmission, places with the most risk of infection, types of agents, attributes and variables that must be added, for obtaining the model of contact propagation of COVID-19. As result, the model has been enhanced by including new agents, and by adding new variables and new behaviors in all the agents that participate in the transmission process.

Future work consists of the Computational model design for the "Covid 19 airborne transmission" and its integration in the ED Simulator. The next step will be the execution, validation, and verification stages for improving the model. The computational model will allow the ED managers to analyze and evaluate potential solutions (the possible effects of applying the control policies' measurements) for propagation of nosocomial infection in a virtual environment and to evaluate the effectiveness of different combinations of laboratory tests, isolation, and other control policies, with the purpose of identifying the best infection control policy. The final step would be the simulation of the Covid 19 transmission in the ED: The "touch" and "airborne" transmissions models approaches. It would be a very interesting idea to apply both approach simultaneously and study how the system is responding.

Acknowledgments. This research has been supported by the Agencia Estatal de Investigacion (AEI), Spain and the Fondo Europeo de Desarrollo Regional (FEDER) UE, under contract PID2020-112496GB-I00 and partially funded by the Fundacion Escuelas Universitarias Gimbernat (EUG).

References

1. Davoli, E., World Health Organization: A practical tool for the preparation of a hospital crisis preparedness plan, with special focus on pandemic influenza. WHO Regional Office for Europe, No. EUR/06/5064207, Copenhagen (2006)
2. Taboada, M., Cabrera, E., Iglesias, M.L., Epelde, F., Luque, E.: An agent-based decision support system for hospitals emergency departments. Procedia Comput. Sci. **4**, 1870–1879 (2011)
3. Cabrera, E., Luque, E., Taboada, M., Epelde, F., Iglesias, M.L.: ABMS optimization for emergency departments. In: Proceedings of the 2012 Winter Simulation Conference (WSC), pp. 1–12. IEEE (2012)
4. Liu, Z., Rexachs, D., Luque, E., Epelde, F., Cabrera, E.: Simulating the micro-level behavior of emergency department for macro-level features prediction. In: 2015 Winter Simulation Conference (WSC), pp. 171–182. IEEE (2015)
5. Jaramillo, C., Rexachs, D., Luque, E., Epelde, F., Taboada, M.: Modeling the contact propagation of nosocomial infection in hospital emergency departments. In: The Sixth International Conference on Advances in System Simulation, pp. 84–89. SIMUL (2014)
6. Wang, Y., Xiong, H., Liu, S., Jung, A., Stone, T., Chukoskie, L.: Simulation agent-based model to demonstrate the transmission of COVID-19 and effectiveness of different public health strategies. Front. Comput. Sci. **82**, 1–8 (2021)
7. Hinch, R., et al.: OpenABM-Covid19—an agent-based model for non-pharmaceutical interventions against COVID-19 including contact tracing. PLoS Comput. Biol. **17**(7), e1009146 (2021)
8. Ponsford, L.R., Weaver, M.A., Potter, M.: Best practices identified in an academic hospital emergency department to reduce transmission of COVID-19. Adv. Emerg. Nurs. J. **43**(4), 355 (2021)
9. Howick, S., McLafferty, D., Anderson, G.H., Pravinkumar, S.J., Van Der Meer, R., Megiddo, I.: Evaluating intervention strategies in controlling coronavirus disease 2019 (COVID-19) spread in care homes: An agent-based model. Infect. Control Hosp. Epidemiol. **42**(9), 1060–1070 (2021)
10. Morawska, L., Milton, D.K.: It is time to address airborne transmission of COVID-19. Clin. Infect. Dis. **71**(9), 2311–2313 (2020)
11. Morawska, L., et al.: How can airborne transmission of COVID-19 indoors be minimised? Environ. Int. **142**, 105832 (2020)
12. Rahaman, H., Barik, D.: Investigation of airborne spread of COVID-19 using a hybrid agent-based model: a case study of the UK. R. Soc. Open Sci. **10**(7), 230377 (2023)

Emerging Topics

QuantumUnit: A Proposal for Classic Multi-qubit Assertion Development

Ignacio García-Rodríguez de Guzmán[(✉)], Antonio García de la Barrera Amo, Manuel Ángel Serrano, Macario Polo, and Mario Piattini

Instituto de Tecnologías y Sistemas de Información, Universidad de Castilla-La Mancha, Camino de Moledores, s/n, 13071 Ciudad Real, Spain
{Ignacio.GRodriguez,Yolanda.Galvan,Antonio.GAmo,Manuel.Serrano, Macario.Polo,Mario.Piattini}@uclm.es

Abstract. In this work, we present a systematic approach to implementing basic assertions in quantum circuits in order to verify classical states across any number of qubits. Our methodology utilizes fundamental quantum gates available on all gate-based quantum computing platforms. Through strategic combinations of these gates, we construct the 'Equal to', 'Different than', 'Greater than', and 'Lower than' comparators. By introducing single-qubit comparators and demonstrating how they can be expanded to handle multiple qubits, we provide a scalable method for verifying classical states in quantum circuits. This work provides the groundwork for robust testing procedures in quantum computing, a critical factor in the continued growth and reliability of this rapidly advancing field.

Keywords: quantum computing · quantum testing · quantum assertion · quantum software engineering

1 Introduction

Quantum computing, a nascent field leveraging the principles of quantum mechanics, promises to revolutionize our computational capabilities by solving certain problems exponentially faster than classical computers. As these quantum systems grow in complexity and potential, the need for rigorous testing mechanisms to ensure their quality becomes increasingly critical [10], considering, in addition, the serious problem of the lack of experience of today's software engineers in quantum software development [9]. In [11], authors focus on some Quantum Software Engineering (QSE) areas which are, according to the current quantum software technology context, the most important to be developed to ensure quantum software sustainability and in turn, quantum computing future: (i) software design of quantum hybrid systems, (ii) testing techniques for quantum programs, (iii) quantum programs quality, and (iv) re-engineering and modernization toward classical-quantum information systems.

In this work, the main concern is testing and quality of quantum software considering that the "zero defect challenge" is also addressed at the Talavera Manifesto [10], and the current state of the art at QSE is in a very early stage. So, with this urgency in mind, the concept of *"assertion"*, widely used in the context of (classical) software testing [13] is considered. An assertion, is *"a boolean expression that is placed at a certain point in a program to check its behaviour at runtime"* [2]. The concept of *assertion* seems to be very suitable for quantum software, not only for the previous success in classical software testing, but also for its applicability to environments where it is necessary to address uncertainty in output, as [13] states. Therefore, we present a general method that allow us to implement basic scalable assertions for verifying classical states (of any number of qubits) in quantum circuits. These methods will provide essential tools for validating and optimizing quantum algorithms, contributing to ensure quantum software quality.

This paper is organized as follows: Sect. 2 develops the context of the work, explaining various concepts related to quantum circuits, necessary to properly understand the proposal, as well as concepts related to quantum testing in general, and property-based testing; Sect. 3, presents the set of multi-cubit assumptions for quantum circuits, which is the basis of this proposal; finally, Sect. 4 develops some conclusions and future work.

2 Context

Quantum computing is a rapidly advancing field that leverages the principles of quantum mechanics to process information. Unlike classical computers, which store data in binary bits (0 or 1), quantum computers use quantum bits or 'qubits' that can exist in multiple states at once due to superposition, enabling parallel computation. Additionally, the phenomenon of entanglement allows qubits that are entangled to be instantaneously connected regardless of the distance separating them, leading to potential increases in computational speed and efficiency.

In this paper, we focus on gate-based quantum computing, a model that manipulates qubits through sequences of quantum gates.

2.1 Quantum Gates

Quantum gates are the basic units of quantum processing. They are the quantum counterpart to classical logic gates and are used to perform operations on qubits. Here, we introduce the gates that are essential for our discussion:

CNOT Gate
The CNOT gate, also known as the Controlled X gate, is a fundamental building block in quantum computing. Its function is to flip the state of a target qubit if the control qubit is in the state |1⟩. This gate provides the basis for the construction of more complex quantum circuits, playing a crucial role in entangling qubits, a fundamental feature of quantum computing.

X Gate
The X gate, also known as the Pauli-X gate, is the quantum equivalent of the NOT gate in classical computing. It flips the state of a qubit, turning $|0\rangle$ into $|1\rangle$, and vice versa.

H (Hadamard) Gate
The Hadamard gate is another essential quantum gate. It transforms the basis states $|0\rangle$ and $|1\rangle$ into superposition states, playing a vital role in creating superposition in quantum circuits.

T Gate
The T gate, or $\pi/8$ gate, applies a phase of $e^{i\pi/4}$ to the state $|1\rangle$. This gate is significant for quantum error correction and the creation of certain quantum states.

2.2 Quantum Software Testing

Quantum software testing, although not yet a mature field of study, has seen significant advances in adapting classical testing methods to the quantum domain. Currently, three main groups of approaches can be identified in the literature: probabilistic proofs, applications of formal logic and reversibility-based techniques [4].

Probabilistic Testing
Probabilistic testing arises due to the intrinsic nature of quantum computers, which offer probabilistic measurements when performing classical observations on cubic registers. This feature differs from the deterministic behavior observed in classical computers. Some works in the literature address this "uncertainty monitoring" [5] by estimating the probability of failure, while others use statistically based assertions to perform error detection.

Applications of the Reversibility of Quantum Circuits
As for the applications of circuit reversibility to testing, they are based on the conservation of energy and thus of information [3]. This property gives rise to interesting features from the verification point of view, such as the significant simplification of the test set generation problem [8]. These properties have led to the development of several applications of reversibility in quantum circuit verification [7].

Despite advances in formal logic and in the exploitation of reversibility properties, the main set of approaches to proof implementation are those based on statistical techniques. However, one of the main obstacles to the adoption of these approaches is the strong dependence on simulation implemented on classical computers, whose scalability is limited due to the difficulty of simulating quantum phenomena. Therefore, it is necessary to develop testing techniques that operate directly on real quantum computing platforms.

Property-Based Testing and Assertions
Property-based testing is a method used in software testing where properties, or specific traits, are defined that the software under test must satisfy. These properties are then used to generate test cases. This differs from traditional example-based testing where individual examples of inputs and their expected outputs are given.

A key element in our approach is the concept of assertions in quantum circuits. In software testing, an assertion is a statement that is assumed to be true at the point they are made within a program. Assertions serve as checkpoints and allow the system to verify that the quantum state is as expected. When an assertion fails, it usually means there's an error somewhere in the code that needs to be fixed. There are currently several approaches for the assertion of properties in quantum programs [6, 14]. In our case, we use assertions to test whether certain conditions or properties hold in our quantum circuits through additional logic added on the same level of the platform.

3 Quantum Multi-qubit Classical Value Assertions Building

In this section, we will illustrate the process of implementing comparators, starting from quantum gates that are common to all quantum platforms. We will build up from these basic quantum gates iteratively to reach our desired assertions.

3.1 First Level: Basic Building Blocks

At this level, we start with the fundamental quantum gate: the Controlled NOT (CNOT) gate. Through a series of manipulations, we expand its capability to handle multiple control qubits, thereby enhancing its functionality.

Reverse CNOT Controls
Certain notations for describing quantum circuits allow the use of control qubits that trigger on a classical value of $|0\rangle$ instead of $|1\rangle$. This capability can be quite useful when designing certain types of quantum circuits. However, when this tool is not available, it becomes necessary to compose the reversed controls by using standard CNOT gates plus other gates: this is achieved by first applying a NOT ("X") gate to the control qubits, inverting their values, then using standard CNOT gate, and finally reapplying the NOT gate to return the control qubits to their original state, as illustrated in Fig. 1.

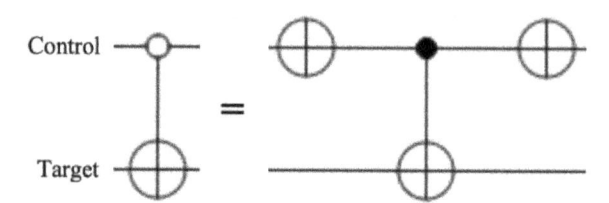

Fig. 1. Example implementation of a CNOT gate with reversed control

The Toffoli Gate

The Toffoli gate, also known as the CCNOT (Controlled-Controlled NOT) gate, is similar to the NAND gate in classical computation. The Toffoli gate operates on three qubits, two of which are control qubits, and the third is the target bit. If the first two qubits are set (both are 1), it inverts (or flips) the third qubit. An implementation example is depicted in Fig. 2.

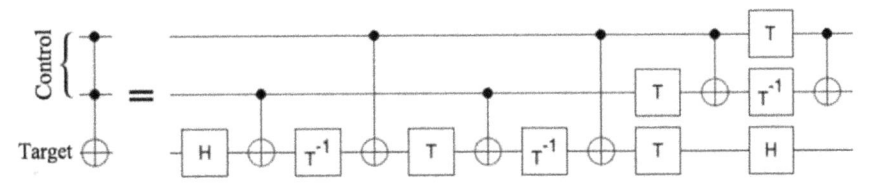

Fig. 2. Implementation example of a Toffoli gate

Multicontrolled Tofolli

A gate can be "*multicontrolled*", meaning there can be more than two control qubits that determine whether the operation will be executed. Some languages have constructs to do this directly, e.g., Qiskit, has the MCXGate (for multi-controlled X gate).

For an "*n-controlled*" gate, you need to use the n control qubits, one target qubit, and $n - 1$ ancilla (helper) qubits (see Fig. 3).

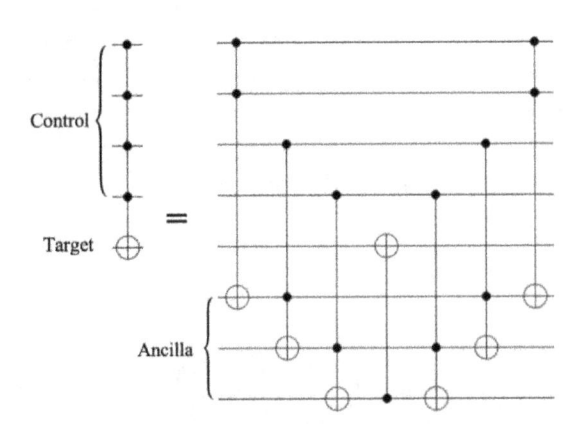

Fig. 3. Implementation of a 4-qubit multicontrolled X gate (CCCCNOT)

3.2 Second Level: Single-Qubit Comparators

With these building blocks, it's possible to create small oracles that implement equal to and different assertions to compare the classical states of qubits than comparators using Toffoli gates and the trick to invert the controls.

Single-Qubit 'Equal Than'

The 'Equal than' comparator can be implemented by placing two Toffoli gates, one normal and the other with inverted controls, as illustrated in Fig. 4. The first will negate the target qubit if the two control qubits are both in the state |1⟩. The second will do the same if both control qubits are in the state |0⟩. In this way, after the execution of the oracle, the target qubit will have been negated if and only if the two control qubits are in the same classical state.

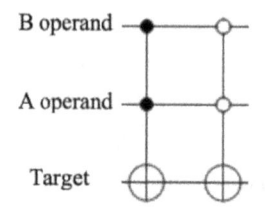

Fig. 4. Implementation of a single-qubit 'Equal than' oracle

Single-Qubit 'Different Than'

Figure 5 shows how to implement a 'Different than' comparator can be implemented by placing two Toffoli gates, one with the first control inverted and the other with the second control inverted. The first will negate the target qubit if the control qubits are in the classical states |0⟩ and |1⟩, respectively (i.e. the qubits are not in a superposition state). The second will do the same if the control qubits are in the states |1⟩ and |0⟩. In this way, after the execution of the oracle, the target qubit will have been negated if and only if the two control qubits are not in the same classical state.

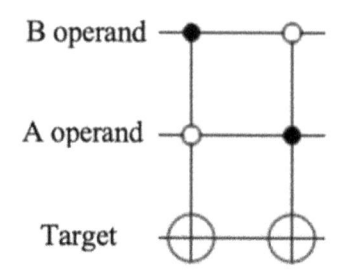

Fig. 5. Implementation of a single-qubit 'Different than' oracle

3.3 Third Level: Multiqubit Comparators

Once we have our single-qubit comparators, we can use them to construct scalable assertions for a greater number of qubits (i.e. that the construction of the assertions can be scaled to the number of qubits in the circuit). Below, the procedure to construct 'Equal than', 'Different than', 'Greater than', 'Lower than', and 'Between' assertions for any number of qubits will be explained.

Scalable 'Equal to' Assertion

To construct this assertion, we need to use single-qubit 'Equal to' operations to bitwise compare the two registers, and then use a multicontrolled Toffoli to verify that all comparisons were positive [1]. For an n-qubit operation, we will need $3n + 1$ qubits, as illustrated in Fig. 6.

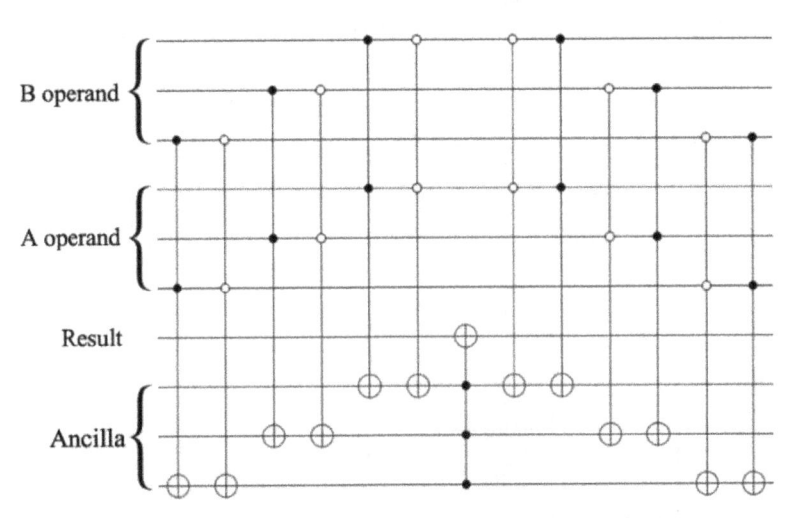

Fig. 6. Composition of the scalable 'Equal than' assertion [1]

Scalable 'Different Than' Assertion

Analogous to classical computation, to construct the inverse of the 'Equal to' assertion, we only need to negate the result. This can be achieved by applying a single-qubit NOT gate to the result qubit after the 'Equal to' operation, (see Fig. 7).

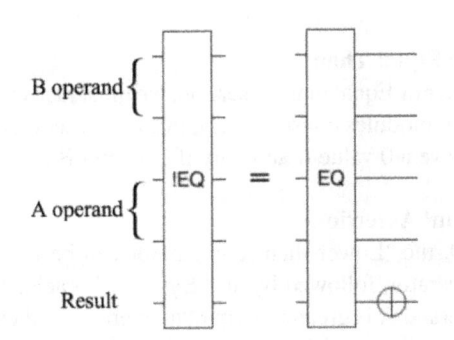

Fig. 7. Composition of the 'Different than' assertion

Scalable 'Greater Than' Assertion

This returns |1⟩ if the more significant register is greater than the less significant one, and |0⟩ if it is less than or equal.

To construct this oracle, we need $3n + 1$ qubits, where n is the size of the registers to compare (refer to Fig. 8). We arrange "$n + n$" qubits for the registers, one qubit for the result, and n ancilla qubits. We use single-qubit *"Different than"* operations, starting from the most significant qubit (the qubit which represent the highest bit of the classical value denoted by the quantum register) and dumping the result in the most significant qubit of the ancilla register. In case both the ancilla qubit and the most significant qubit of the more significant register are in the state |1⟩, we negate the result qubit.

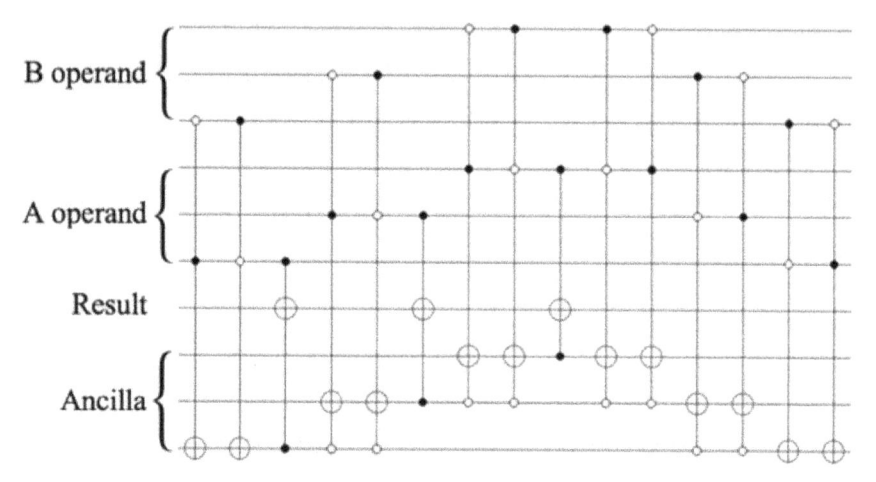

Fig. 8. Composition of a scalable 'Greater than' assertion

At this point, the result qubit will be 1 if the A operand is greater than the B operand from the first digit. Then subsequently we have to do the same operation to the less significant ones, this time also negative controlling that the previous comparisons weren't positive. In this way, the result qubit has a |1⟩ value if and only if the A operand is greater than the B operand.

Scalable 'Greater or Equal Than'

To implement a 'Greater or Equal than' assertion, we must just simply append a 'Greater than' and an 'Equal to' modules consecutively, in such a way that, after their execution, the target qubit will have a 0 value if and only if Operand B is lower than Operand A.

Scalable 'Lower Than' Assertion

As depicted in Fig. 9, the 'Lower than' comparator can be formed by first using the *"Greater than"* comparator, followed by an *"Equal to"* oracle. At this point the target qubit will be |1⟩ if operand A is greater or equal to operand B, therefore we only need to negate the target qubit to have our *"Lower than"* oracle. This can be done by applying a single-qubit NOT gate to the result qubit after the 'Greater than' and 'Equal to' operations. It is interesting to note that there are approaches to the automatic generation of scalable 'Less than' oracles for quantum amplitude amplification [12].

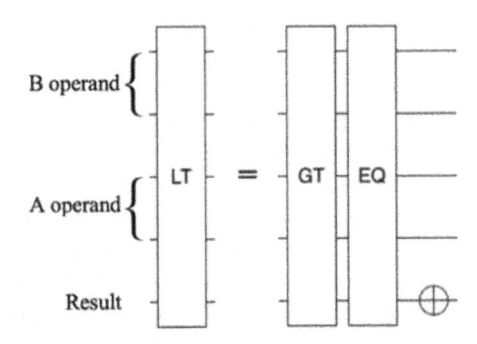

Fig. 9. Implementation example of the 'Lower than' assertion

Scalable 'Lower Than' Assertion

Analogously, we can implement an 'Lower or Equal than' assertion by appending a 'Lower than' and a 'Equal to' oracles.

Scalable 'Between' Assertion

The 'Between' assertion is an aggregation of two consecutive 'Greater than' assertions. The first one will check that operand A is greater than operand B, the second one will check that operand C is greater than operand A. Finally, we need to add a Toffoli gate to check that both the previous assertions has been positive. To accomplish this, we need two additional ancilla qubits, for a total of 4*n + 3 qubits where n is the size of the operands (n for coding each of the three operands, n + 2 ancilla and the result qubit). Refer to Fig. 10. For a detailed representation.

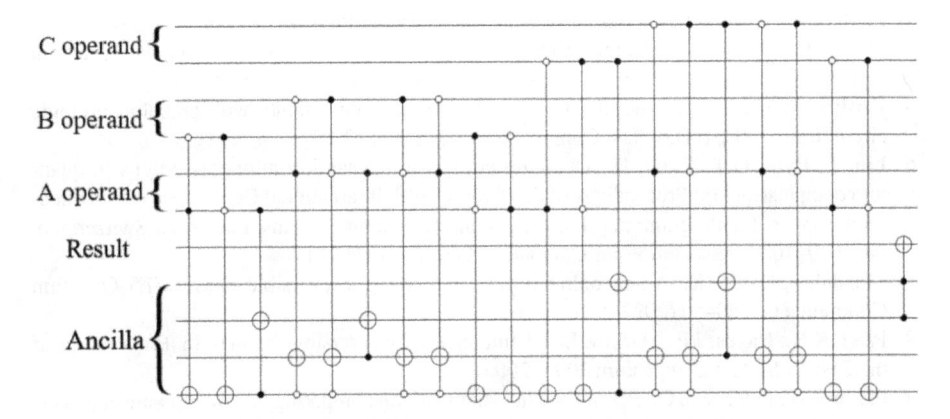

Fig. 10. Detail of the proposed implementation of the 'Between' assertion

4 Conclusions

In this work, we have established a novel framework for implementing basic assertions in quantum circuits that can verify classical states of any number of qubits. Our approach, which uses a layered, iterative structure, is grounded in universal quantum gates such as the CNOT, H, and T gates that are available on any gate-based quantum computing platform. By combining these basic elements in a systematic manner, we have demonstrated the construction of 'Equal to', 'Different than', 'Greater than', 'Lower than', 'Greater or Equal than', 'Lower or Equal than' and 'Between' comparators, forming a basic toolbox for testing quantum algorithms.

The work presented here is a practical step towards creating more systematic and measurable development tools and procedures for quantum computing. It underscores the importance of structured, replicable processes in the development of quantum algorithms and architectures. As we continue to make progress in the field, these methodical approaches will help improve the reliability and scalability of quantum technologies. Future research will build on this work, refining these tools and procedures as we deepen our understanding of quantum computing's capabilities and challenges.

Acknowledgements. This work has been carried out in the context of the projects QSERVU-CLM (PID2021-124054OB-C32) financed by the Spanish Ministry of Science and Innovation (MICINN) and the European Union, and the financial support for the execution of applied research projects, within the framework of the UCLM Own Research Plan, co-financed at 85% by the European Regional Development Fund (FEDER) UNION (2022-GRIN-34110).

References

1. Amo, A., Serrano, M., Guzmán, I., Usaola, M., Piattini, M.:Automatic generation of testing circuits for deterministic quantum algorithms (2023)
2. Barr, E.T., Harman, M., McMinn, P., Shahbaz, M., Yoo, S.: The oracle problem in software testing: a survey. IEEE Trans. Softw. Eng. **41**(5), 507–525 (2015)
3. Fredkin, E., Toffoli, T.: Conservative logic. Int. J. Theor. Phys. **21**(3), 219–253 (1982)
4. García de la Barrera, A., García-Rodríguez de Guzmán, I., Polo, M., Piattini, M.: Quantum software testing: state of the art. J. Softw. Evol. Process **35**(4), e2419 (2023)
5. Krishnaswamy, S., Markov, I.L., Hayes, J.P.: Tracking uncertainty with probabilistic logic circuit testing. IEEE Des. Test Comput. **24**(4), 312–321 (2007)
6. Liu, J., Byrd, G.T., Zhou, H.: Quantum circuits for dynamic runtime assertions in quantum computation. In: Proceedings of the Twenty-Fifth International Conference on Architectural Support for Programming Languages and Operating Systems, Lausanne, Switzerland, pp. 1017–1030. Association for Computing Machinery (2020)
7. Mondal, J., Das, D.K.: A new online testing technique for reversible circuits. IET Quantum Commun. **3**(1), 50–59 (2022)
8. Patel, K.N., Hayes, J.P., Markov, I.L.: Fault testing for reversible circuits. In: Proceedings of the 21st VLSI Test Symposium 2003 (2003)
9. Piattini, M., Peterssen, G., Pérez-Castillo, R.: Quantum computing: a new software engineering golden age. ACM SIGSOFT Softw. Eng. Notes **45**(3), 12–14 (2020)
10. Piattini, M., et al.: The Talavera manifesto for quantum software engineering and programming (2020)

11. Piattini, M., Serrano, M., Perez-Castillo, R., Petersen, G., Hevia, J.L.: Toward a quantum software engineering. IT Prof. **23**(1), 62–66 (2021)
12. Sanchez-Rivero, J., Talavan, D., Garcia-Alonso, J., Ruiz-Cortes, A., Murillo, J.: Automatic generation of an efficient less-than oracle for quantum amplitude amplification. In: 2023 IEEE/ACM 4th International Workshop on Quantum Software Engineering (Q-SE) (2023)
13. Taromirad, M., Runeson, P.: A literature survey of assertions in software testing, pp. 75–96 (2024)
14. Ying, M.: Birkhoff-von Neumann quantum logic as an assertion language for quantum programs (2022)

Tool for Quantum-Classical Software Lifecycle

Jesús Párraga Aranda[1]([✉]), Ricardo Pérez del Castillo[2], and Mario Piattini[2]

[1] Universidad Pontificia de Salamanca, Salamanca, Spain
jesus.parraga.aranda@gmail.com
[2] Universidad de Castilla la Mancha, Toledo, Spain
{ricardo.pdelcastillo,mario.piattini}@uclm.es

Abstract. With the growing diffusion of quantum computing technology and the increasingly promising applications derived from it, the relevance of developing specific software for these systems is gaining significant momentum. This surge is due to the need to design and produce quantum software that meets performance and functional requirements but also follows the well-known good practices and rigorous methodologies inherent in quantum software engineering. In this context, one of the main challenges facing the development of hybrid (quantum-classical) systems is the effective management of the lifecycle of this new type of software, whose nature differs from traditional systems. The proposed research attempts to comprehensively address the lifecycle management of hybrid software, through the design and development of a specific support tool. To achieve this goal, the ICSM (Integrated Software Cycle Management) model, which is a consolidated framework for the lifecycle management of traditional software, will be taken as a starting point. This model will be carefully adapted to meet the unique needs and challenges inherent in hybrid software, thus ensuring that the development, maintenance, and updating practices of this type of software are as robust and efficient as those applied in the realm of conventional software. Through this adaptation, the aim is not only to improve the quality of the developed hybrid software but also make developers easier to adopt the innovative and complex quantum software paradigm.

Keywords: Quantum Computing · Quantum Software Engineering · Hybrid software · Software Lifecycle · ICSM

1 Introduction

Quantum computing inaugurates a new era for software engineering [1], and we are witnessing research proposing different methodologies, life cycles, techniques, and tools to build a true quantum software engineering (QSE) discipline [2–4].

However, for quantum applications to be truly useful and not just PoCs (Proof of Concepts), they must be developed according to the best practices of Software Engineering [5] to ensure their quality.

Quantum computing introduces concepts and operational principles radically different from classical computing, such as superposition and quantum entanglement [6]. This inherent complexity demands a specialized approach to project management that can address the technical and theoretical peculiarities of quantum software development. Since we are currently in a transitional phase towards quantum computing, most practical applications will involve hybrid systems integrating quantum and classical components [7]. Managing the lifecycle of these hybrid systems requires a tool capable of handling the complexity of both worlds, facilitating effective and efficient integration.

In recent years, various tools have been proposed for the development of quantum systems [3, 8]. However, none are specialized in lifecycle management based on a risk-driven model, like ICSM.

2 Characteristics of Hybrid Lifecycle

The development of quantum-classical software presents unique challenges that require a specialized approach to its lifecycle. This type of software seeks to leverage the advantages of quantum computing, such as its potential to solve certain problems much faster than classical computers, while maintaining the compatibility and functionality offered by classical systems.

The key characteristics of the quantum-classical software lifecycle are as follows:

- Hybrid development phases (quantum-classical).
- Iterative and Flexible.
- Integrated Testing.
- Dependency on specialized infrastructure.
- Multidisciplinary Collaboration.
- Management of complexity and uncertainty.
- Adaptability to emerging standards.

There are multiple lifecycle proposals for quantum software development:

- In [9], a lifecycle for quantum development (QDLC) inspired by the waterfall model is proposed.
- In [10], another model is provided, based on "quantum data provenance." This provenance data will serve different phases of the lifecycle such as error analysis or quantum hardware selection. This model starts with the quantum/classical division phase and ends with results analysis, although it only focuses on circuit-based quantum systems, not considering annealing solutions.
- In [11], an "interactive advisor" is presented, which guides the decision on what type of architecture to implement: classical, hybrid, or quantum.
- In [12], the aQuantum hybrid lifecycle supported by the QuantumPath® tool is proposed, which, through iterative phases, allows the integration of quantum systems (both gate-based and annealing) using qSOA® [13].

Nevertheless, risk management is a critical facet in the software development lifecycle, especially in the development of classical-quantum software, which is hardly ever considered in a holistic way. Therefore, we believe that the ICSM model [14] is the best foundation for implementing a lifecycle for hybrid (quantum-classical) systems.

3 ICSM – Incremental Commitment Spiral

The ICSM model consists of two main stages: incremental definition and incremental development and production operations [14]. As shown in Fig. 1, the model is based on a spiral approach and is founded on the following elements:

- Value-driven approach by stakeholders.
- Commitment and responsibility that progressively intensify.
- Simultaneous application of multidisciplinary engineering.
- Decision-making based on risk assessments and evidence.

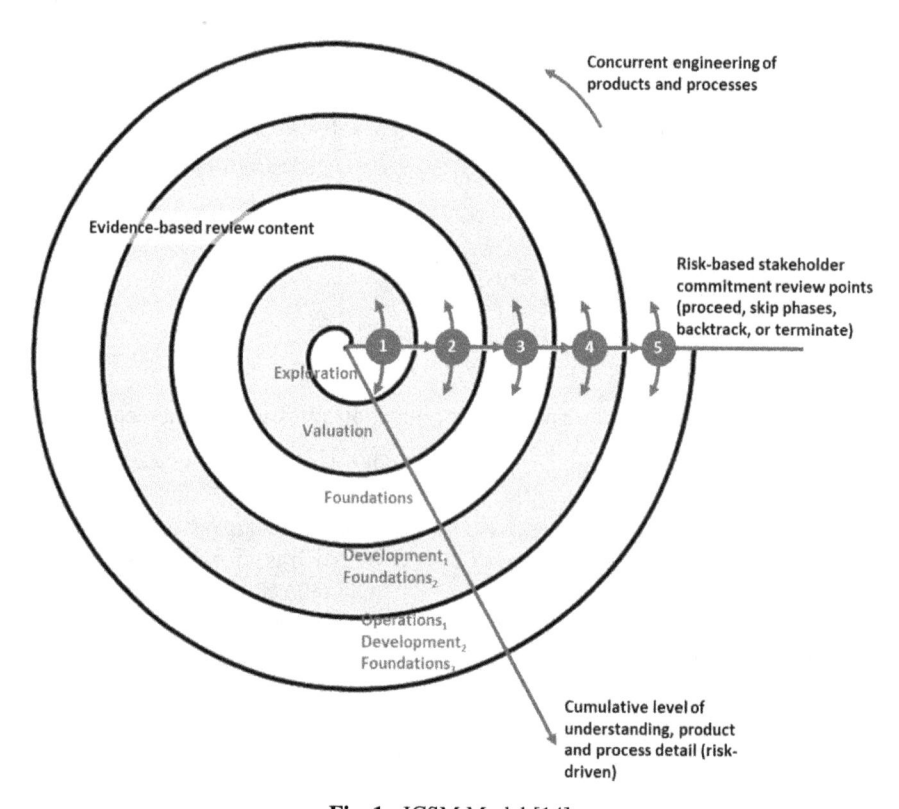

Fig. 1. ICSM Model [14]

The lifecycle progresses based on the risk-based decision-making process. If the risk is manageable, the project advances to the next spiral phase. In cases where the risks are high but controllable, the project remains in the current phase until these risks are mitigated through effective mitigation plans. If the risks are minimal, it is possible to group spirals. Conversely, if the risks are excessively high or unmanageable, the project could be terminated or replanned. Based on the above and as can be seen in [15], we can observe how ICSM adapts to quantum development (see Fig. 2).

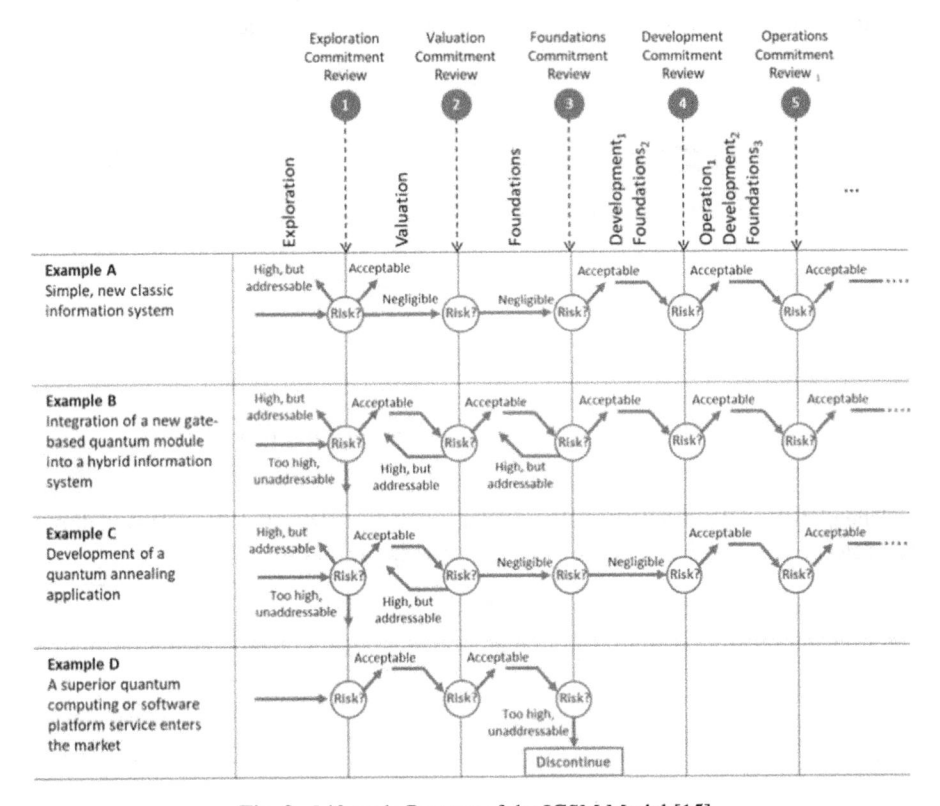

Fig. 2. Lifecycle Process of the ICSM Model [15]

4 QLCManager

In this paper, we present QLCManager (Quantum Life Cycle Manager), an application that offers a comprehensive solution specifically designed to facilitate the effective management of hybrid projects that implement the ICSM model adapted to hybrid (quantum-classical) development. This robust and user-friendly platform provides different users with the necessary tools to plan, execute, and monitor project progress, ensuring that each phase of the lifecycle is managed optimally.

This tool is designed to perfectly adhere to the phases of the Incremental Commitment Spiral Model (ICSM), allowing users to identify potential risks at different stages of the project, analyze their possible impact and probability, and develop effective mitigation plans. The application facilitates the continuous updating of risk information and their reevaluation as the project progresses and evolves, ensuring that risk management is proactive and adaptive. The planning and monitoring of iterations in QLCManager enable teams to structure and closely follow each spiral of the ICSM.

The list of projects to which users have access is presented as in Fig. 3, where it is easy to see in which phase of the spiral (as provided by ICSM) each project is currently situated.

Fig. 3. Project States Categorization base on their current phase

The phases of the projects and the formation of the spiral at each step taken during the project's lifecycle are presented to the user as shown in Fig. 4.

Fig. 4. Project phases

4.1 Risks Creation

QLCManager offers a list of predefined risks (See Fig. 5).

Stage	Phase	Key Decision Inputs & Risks Assessment
Definition	Exploration	Solve the problem with classical software
		Solve the problem with some quantum software components
		Existing algorithms to used/adapted
		New algorithms to be developed
	Valuation	Existing information system to be re-engineered (totally or partially) to quantum paradigm
		Specific quantum toolkits for specific domains/sectors
		Needs for workflow operations between classical and quantum
Development Operations & Production	Foundation	Service-Orientation challenges: service composition, configuration management, monitoring, security
		Exploration using simulators instead of actual quantum computers (which are more expensive)
		Error tolerance and power (e.g., number of qubits) of quantum hardware
		Quantum technology stability
		Available quantum software workforce
	Development	Co-design quantum software-hardware
		Unitary testing for quantum software
		Integration testing for quantum algorithms and classical software drivers
		New opportunities can suggest shifting some classical functionalities to quantum
		Volatility of the today's quantum software development technology
	Operation	Error tolerance and power (e.g., number of qubits) of quantum hardware
		Scalability (requests, users, etc.)

Fig. 5. Common decisions during risk Assessment

In addition to those risks mentioned above, risks can be created from scratch or a pre-existing risk in the system can be selected, whether from the current project or any other accessible project by user (see Fig. 6).

Fig. 6. Add risk to the current phase action.

4.2 Validation Phase

Phase validation (see Fig. 7) allows the system to automatically offer the best decision for the project and phase concerning the context of risks, their criticality, and the number of those remaining to be mitigated.

The system's automatic validation to propose the decision based on the context is as follows:

- Blocking level risks, 1 or more

  - Phase evaluation: Inaccessible
  - Proposal: Discontinue the Project

- High-level risks, 5 or more

  - Phase evaluation: Addressable
  - Proposal: Go Back

- Medium-level risks, 5 or more

– Phase evaluation: Acceptable
  – Proposal: Move Forward

- Low-level risks do not influence the decision-making for phase evaluation and the final proposal.

The system will provide an action proposal, but eventually, is the user, who knows the project context, which make the final decision.

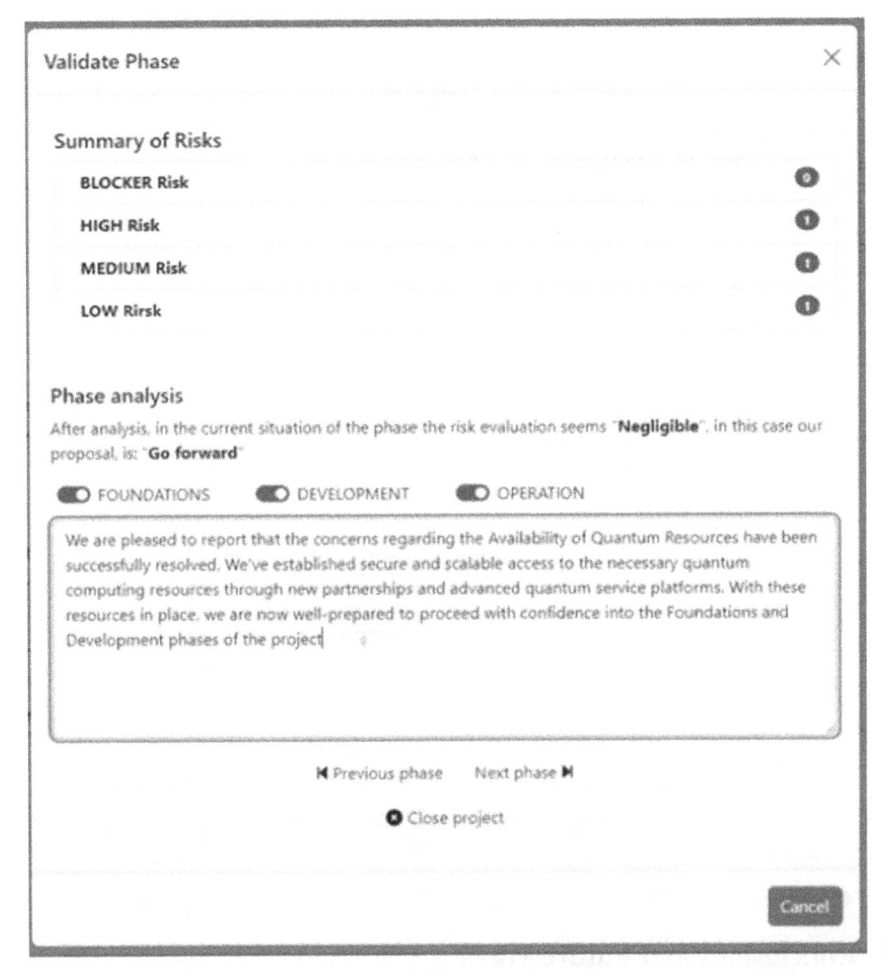

Fig. 7. Validation phase

4.3 Timeline

QLCManager provides a graphical representation of the project timeline, Fig. 8, where all the decisions made during the project's lifecycle are represented. Each element of the timeline contains the necessary information to maintain and visualize the trace of the project's evolution. We highlight the following attributes:

- Iteration number
- Directionality between phases, Previous phase >/< New phase
- Phase evaluation result
- Decision proposed by the system.
- Comments specified by the user when performing phase evaluation and decision-making.
- Notification if the user made a different decision than the one proposed by the system's automatic evaluation "!".

The timeline is organized as follows, if the card is on the right side of the timeline, it means that the project has moved "forward" within the spiral of the lifecycle. Conversely, if it is on the left side, it indicates a "backward step" between phases.

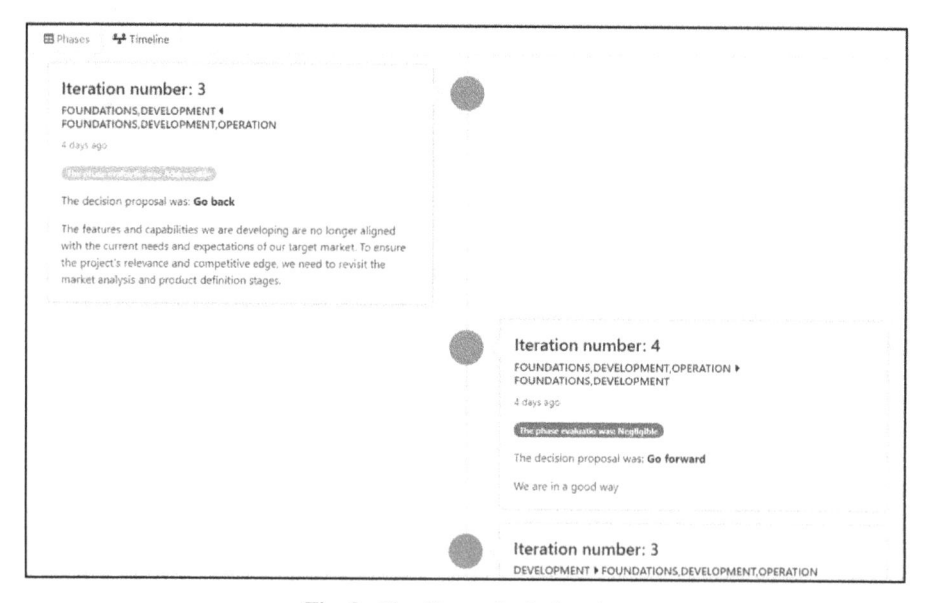

Fig. 8. Timeline and spiral cycles

5 Conclusions and Future Work

One of the main factors that will ensure the success of quantum software and potentially usher in a new golden age of software engineering [1] is not forgetting its principles during development [5].

After an exhaustive analysis to identify the most appropriate lifecycle that aligns with the demands of a hybrid software project, it was concluded that the Incremental

Commitment Spiral Model (ICSM) is a suitable option due to the nature of hybrid software development. Consequently, work has been done to adapt it to hybrid development [15].

Based on this proposal and given the lack of specific tools in the market to manage such projects, the QLCManager tool was developed. It offers specialized functionalities for integrated risk management, iterative planning, and effective collaboration among multidisciplinary teams. QLCManager establishes itself as an indispensable solution to improve efficiency and effectiveness in executing complex projects, such as all those of a hybrid nature.

Future work includes the validation of the tool by the community through its use in real projects. Potential extensions to the tool could include incorporating hybrid project management patterns based on their characteristics.

Acknowledgments. This work has been supported by grants PID2022-137944NB-I00 (SMOOTH Project) and PDC2022-133051-I00 (QU-ASAP Project) funded by MICIU/AEI/10.13039/501100011033 and by the European Union NextGenerationEU/ PRTR and UCLM Own Research Plan, co-financed at 85% by the European Regional Development Fund (FEDER) UNION (2022-GRIN-34110).

References

1. Piattini, M., Peterssen, G., Pérez-Castillo, R.: Quantum computing: a new software engineering golden age. ACM SIGSOFT Softw. Eng. Notes **45**(3), 12–14 (2020). https://dl.acm.org/doi/10.1145/3402127.3402131
2. Zhao, J.: Quantum software engineering: landscapes and horizons. arXiv preprint arXiv:2007.07047 (2020)
3. Hevia, J.L., Peterssen, G., Ebert, C., Piattini, M.: Quantum computing. IEEE Softw. **38**(5), 7–15 (2021)
4. Serrano, M., Pérez, R., Piattini, M. (eds.): Quantum Software Engineering. Springer, Berlin (2022)
5. Piattini, M., Peterssen, G., Pérez-Castillo, R., Hevia, J.L., et al.: The talavera manifesto for quantum software engineering and programming. In: QANSWER 2020 QuANtum SoftWare Engineering & Programming. Proceedings of the 1st International Workshop on the QuANtum SoftWare Engineering & Programming, Talavera de la Reina, Spain, 11–12 February 2020, pp. 1–5 (2020). http://ceur-ws.org/Vol-2561/paper0.pdf
6. Piattini, M., Serrano, M.A., Cruz-Lemus, J.A., Pérez del Castillo, R.: Informática Cuántica. Amazon (2022)
7. Akbar, M.A., Khan, A.A., Rafi, S.: A systematic decision-making framework for tackling quantum software engineering challenges. Autom. Softw. Eng. **30**(2), 22 (2023)
8. Serrano, M., Cruz-Lemus, J.A., Pérez-Castillo, R., Piattini, M.: Quantum software components and platforms: overview and quality assessment. ACM Comput. Surv. **55**(8), 164:1-164:31 (2023)
9. Dey, N., Ghosh, M., Samir, S., Chakrabarti, A.: QDLC - the quantum development life cycle. 2010.08053v1 [cs.ET] (2020)
10. Weder, B., Barzen, J., Leymann, F., Salm, M., Vietz. D.: The quantum software lifecycle. In: Proceedings of the 1st ACM SIGSOFT International Workshop on Architectures and Paradigms for Engineering Quantum Software (APEQS 2020), pp. 2–9. Association for Computing Machinery, New York (2020). https://doi.org/10.1145/3412451.3428497

11. Misra, J., Kaulgud, V., Kaslay, R., Podder, S.: When to build quantum software? arXiv:2104.09117v1 (2021)
12. Hevia, J.L., Peterssen, G., Piattini, M.: QuantumPath: a quantum software development platform. Softw. Pract. Exp. **52**(6), 1517–1530 (2022)
13. Hevia, J.L., Peterssen, G., Piattini, M.: QSOA®: dynamic integration for hybrid quantum/classical software systems. J. Syst. Softw. **214**, 112061 (2024)
14. Boehm, B., Turner, R., Lane, J.A., Koolmanojwong, S.: The Incremental Commitment Spiral Model: Principles and Practices for Successful Systems and Software. Pearson Education (2014)
15. Pérez-Castillo, R.: Guidelines to use the incremental commitment spiral model for developing quantum-classical systems. Quantum Inf. Comput. **24**(1&2), 0071–0088 (2024)

Innovation in Computer Science Education

Strategies to Predict Students' Exam Attendance

Gonzalo L. Villarreal[1,2](✉) and Verónica Artola[3]

[1] Universidad Nacional de La Plata, PREBI-SEDICI, Buenos Aires, Argentina
gonzalo@prebi.unlp.edu.ar
[2] Comisión de Investigaciones Científicas, CESGI, Buenos Aires, Argentina
[3] Universidad Nacional de La Plata and Comisión de Investigaciones Científica, III-LIDI, Buenos Aires, Argentina
vartola@lidi.info.unlp.edu.ar

Abstract. This article presents a study on predicting student attendance to exams in a university setting. The study focused on the Concept of Algorithms, Data, and Programs course, a foundational course in systems bachelor. Two models were constructed: linear regression and polynomial regression of degree 3, aimed to predict the total number of attendees and the number of students who would pass the exam. We built a dataset that included information on student enrollment, previous exam attendance, grades, and other relevant factors. Students were classified into three groups: reduced exam, complete exam with prior attendance, and complete exam without prior attendance. The results showed that the models' predictions were accurate enough, and that they could be used to ensure appropriate classroom occupancy without overcrowding or empty rooms. The models guided the allocation of students, optimizing space utilization while providing available seats for attending students. The study identified opportunities for improvement. One limitation was the assignment of attendance probabilities to achieve the overall predicted attendance. Future work could involve predicting attendance rates for each group individually. Additionally, implementing a classification model to categorise students into pass, fail, insufficient, and non-attendance groups would provide a more comprehensive understanding of student outcomes.

Keywords: regression analysis · attendance prediction · approval prediction · effective resource planning

1 Introduction

In educational institutions, predicting student attendance for exams plays a crucial role in effective planning and resource allocation. Having reliable estimates of attendance enables educators to make informed decisions regarding seating arrangements, printing exam materials, and overall logistics. By leveraging predictive models, we can forecast student attendance with reasonable accuracy and facilitate better preparation for exams.

In this article, we will walk you through an implementation of a linear regression and a polynomial regression model to predict student attendance for exams which offer many benefits, including:

- Resource Planning: by knowing the expected number of students attending an exam, administrators can plan seating arrangements, arrange adequate exam materials, and ensure a smooth experience for both students and staff.
- Timely Communication: educational institutions can inform students about essential exam details, such as exam location, timing, and any specific instructions, well in advance
- Performance Analysis: by comparing attendance rates with exam scores, educational institutions can identify potential correlations and gain insights into factors affecting student success.
- Efficient Resource Utilization: for instance, if a lower-than-expected attendance is forecasted, institutions can consider consolidating examination rooms, saving on logistics costs and reducing the environmental impact associated with exam preparations.

The subsequent sections will outline the steps involved in implementing the regression models, including data preparation, model training, evaluation, and prediction.

2 State of the Art

In recent years, several researchers have explored alternatives to predict student attendance in educational settings. Maud Vissers [1] investigated the probability of predicting class attendance for students' personal development, for professors' preparation and intervention, and to optimise universities' educational programs. The author used Logistic Regression, Random Forest and Naïve Bayes in this study. He found that class attendance can be predicted based on sensor data and education data, and the best performing algorithm was the Random Forest algorithm containing GPS Location data, WiFi Location data, and Class Information data.

Muzaferija et al. [2] focused on the reasons why students' attendance decreased, in order to try to predict when it was going to happen, and act on causing factors in order to prevent it. They built a dataset containing 2nd-year student attendance data from two years, and although the dataset didn't contain all the details about the students and their classes it was enough to extract the patterns of student attendance behavior and create a model to predict it. In their study they found that the machine learning algorithm that created the most accurate model was the C4.5 decision tree algorithm with 77.5% accuracy, followed by a Linear Regression algorithm with 75,37% accuracy. Fernandes et al. [3] introduced a classification model based on Gradient Boosting Machine (GBM), the demographic characteristics of the students and the achievement grades obtained from the in-term activities were taken into consideration. In this study, the authors observed the importance of previous year's achievement scores and attendance data for estimating students' achievement.

M. Yağcı [4] proposed a new model based on machine learning (ML) algorithms to predict the final exam grades of undergraduate students. In this model, he considered the students' midterm exam grades as the source data, combined with Department data and Faculty data for each student. He compared different ML algorithms, including logistic regression, k-nearest neighbor algorithms and random forests, among others. The model achieved a classification accuracy of 70–75%, with 71.7% of accuracy for the Linear Regression based model.

Another ML-based approach was proposed by Rashid et al. [5]. In their work, the authors used ML techniques to predict students' attendance to classes.

Considering different reasons why students skip classes, they built a dataset by collecting 2 years of attendance and they used a variety of machine learning algorithms to predict attendance, including LR, support vector machines, and decision trees. They found that all of the algorithms were able to predict attendance with a high degree of accuracy. The authors suggest that teachers can use ML to identify students who are at risk of missing class, and can then take steps to address the needs of these students. They also suggest that ML could be used to improve the efficiency of teaching. For example, teachers could use ML to predict which students are likely to need extra help on a particular topic and thus create individualised learning plans for them.

Retention prediction studies can also contribute with variables and techniques to predict attendance. Robert D. Reason [6] considered high school performance (high school GPA and SAT/ACT scores) as a variable to predict students' graduation rates. He mentions that students who entered college with a high school GPA were more likely to graduate with a degree in 4 years than students who entered with a low GPA. Similarly, higher SAT scores were also associated with higher graduation rates. However, the author clarifies that the effect size of these variables was relatively small. They only predicted 12% of the variation in retention. Even though we are not predicting students' retention, we also consider the results in the initiation course as an important variable in our study. Credé et al. [7] review the relationship of class attendance with grades and student characteristics, and they found strong relationships with class grades and GPA, which seem to be a better predictor of college grades than any other known predictor of academic performance, including scores on standardised admissions tests such as the SAT, high school GPA, study habits, and study skills.

2.1 About the Course of CADP

The course "Concepts of Algorithms, Data, and Programs" (CADP, 2023) is a first-year subject in the Systems Bachelor and Computer Science Bachelor programs. Students taking this course can be either new students who enrolled in the current year or students who enrolled in previous years but did not pass the subject at the time of enrollment. Incoming students undertake a course called "Problem Expression and Algorithms" (EPA), where they are introduced to the basic computational thinking and programming concepts required to progress in CADP. EPA is a prerequisite for CADP, but it is not an eliminatory course per se. To pass EPA, students must only meet a minimum attendance percentage requirement. However, there is an exam at the end of EPA, and students who pass this exam receive certain benefits as rewards for their CADP coursework. These benefits may include the opportunity to take a reduced exam and a priority at the time of selecting course hours.

Once students have successfully completed EPA, they can proceed to CADP, where they delve deeper into algorithms, data structures, and programming concepts. CADP is a comprehensive course that builds upon the foundations established in EPA. It covers topics such as algorithm analysis, data representation, programming paradigms, and problem-solving strategies. The CADP course is structured into lectures, practical sessions, and assignments. Assignments and projects are designed to reinforce the learned concepts and allow students to apply their knowledge to real-world problems.

Throughout CADP, attendance records are maintained to monitor students' participation and engagement in the course. Only students with a certain percentage of attendance are able to take the exam.

At the end of CADP, students must take a final exam to evaluate their understanding of the subject matter, which assesses their knowledge of concepts covered during the course. Successful completion of the exam is a requirement for obtaining a passing grade in CADP and progressing to the subsequent courses in the curriculum. Students have three opportunities to take this exam. We would like to add that the course delivery for CADP lasts for one semester, and there are course retakes offered in the second semester. However, while the course delivery in the first semester is open to all students (both incoming and returning), the course in the second semester is limited only to those students who have completed the course in the first semester, have met the required attendance to take the exam, have taken the exam, and have received a failing grade. This distinction between students in the first and second semesters results in differences in the total number of students attending each semester, as well as the attendance rate for exams and also the pass rate. It is important to consider these distinctions when analyzing the attendance and pass rates within the context of the course. Understanding the differences in student populations and their characteristics between the two semesters provides valuable insights into the dynamics of student performance and the factors that contribute to success or failure in the course. By acknowledging these variations, educators and administrators can develop targeted strategies and interventions to address the specific challenges faced by students in both the traditional course delivery and the course retake.

3 Methodology

To address the problem of predicting student attendance for an exam, two different models were developed: one based on linear regression (LR) and another using a polynomial regression of degree 3 (PR). These models allowed us to predict both the total number of attendees and the total number of students who would pass the exam. Having two predictive models allows us to compare results between them. Additionally, while the LR model is simpler, the PR model may better capture the statistical variations between the first and second semesters, considering the specific course delivery characteristics described earlier. In the context of the CADP course, where different instructional modes and student cohorts are involved, the polynomial model's ability to account for statistical variations between semesters can be advantageous. It can capture nuances and dynamics specific to each semester, such as the difference in enrollment and attendance rates for the traditional course delivery and the course retake.

We followed a systematic methodology that involved the construction and preparation of a comprehensive dataset, with data for the period 2019–2023 [8]. The dataset encompasses information about students eligible to take the exam, students who actually appeared for the exam, and the number of students who achieved different grades. The grades are classified as "Approved" (an exam that was well done, although it may have had some errors), "Disapproved" (an exam that was incorrect but showed an attempt to solve the problem), and "Insufficient" (a very incomplete exam).

Additionally, the dataset includes information about the academic year, semester, exam number and course modality, categorised as "In-person" (during regular on-campus sessions), "Virtual" (during the COVID pandemic when classes were conducted online), and "Hybrid" (during the transition period after the pandemic, combining in-person and virtual elements). During the COVID pandemic, since both classes and exams were conducted online, students' attendance were higher than usual, although pass rates were a bit lower. Not taking into consideration this parameter may introduce severe distortions in our models.

The steps involved in constructing the dataset and preparing it for the models are as follows:

1. Data Collection: We collected information from various sources, including student records, attendance registers, and exam grading records.
2. Data Cleaning: We carefully cleaned the dataset by removing a few duplicates, but specially completing missing values or inconsistent entries. Additionally, we performed data validation checks to ensure the accuracy and integrity of the data.
3. Feature Engineering: To enhance the predictive power of the model, we performed feature engineering. This involved transforming the categorical variables, such as semester, exam number, and course modality, into numerical representations using appropriate encoding techniques.
4. Linear and Polynomial Regression Models: We implemented a LR model and a PR of degree 3 model. Both models take into account the features derived from the dataset.
5. Model Training and Evaluation: We divided the dataset into training and testing sets to train the polynomial regression model. The model was trained using the training set, and its performance was evaluated on the testing set.
6. Prediction and Analysis: Once the model was trained and evaluated, we utilised it to make predictions on new data. These predictions were then analyzed to gain insights into student attendance patterns and to identify factors that significantly impact attendance.

Based on the predictions obtained from the models, the students were categorised into three groups:

1. Students taking a reduced exam (group Reduced): This group consisted of students who had previously passed the EPA exam. They were eligible to take a shorter version of the exam.
2. Students taking a complete exam and attended previous exam (Group Complete): This group comprised students who were attending the complete exam and had also attended the previous exam.
3. Students taking a complete exam and did not attend previous exam (Group Complete Absent): This group consisted of students who were attending the complete exam but had not attended the previous exam.

Each student was assigned to one of these groups based on their eligibility and attendance history. Using this information, the attendance percentages for each group were estimated in such a way that the total attendance would match the predicted attendance from each model (Table 1).

Table 1. Number of students in each of the three groups, and attendance estimation for both models for the second exam.

Group	Number of students	Linear Regression Attendance Estimation	Polynomial Regression Attendance Estimation
Reduced	262	84,10%	86%
Complete	541	79%	80,10%
Complete Absent	966	11%	11,37%

By incorporating this categorization and adjusting the attendance percentages accordingly, we aimed to align the predicted total attendance with the attendance estimated by the selected model. This approach allowed for a more accurate representation of the different student groups and their corresponding attendance patterns.

3.1 Predicting Students that Will Pass the Exam

Although it was not the purpose of these models to predict the results of the exams, the predictions from the initial regression models were further utilised as an input for a subsequent predictive model to estimate the percentage of students who would pass the subject. To achieve this, the predicted attendance values obtained from the regression models were combined with the previous dataset. This data was used to train new predictive models, again using LR and PR, which aimed to forecast the percentage of students who would successfully pass the subject. Once the predictive models were trained and evaluated for accuracy, it was applied to the current cohort of students to make predictions on the probability of passing the subject.

4 Prediction Results

At the moment of writing these lines, the first exam of the year had already been taken, so the dataset also included attendance and results data for this cohort. Thus, this model has been used to predict attendance and pass rates for the second and third exam (in the later we have also included data from the former into the dataset).

For the second exam, there were a total of 14 classrooms available, each with a different maximum capacity. These classrooms were organised into three groups, corresponding to the same groups into which the students were classified (reduced, complete with previous attendance, and complete without previous attendance). Next, the students were assigned to the classrooms based on their respective groups. The goal was to achieve an occupancy level between 40 and 90 percent in each classroom. This approach aimed to maximise the utilization of available space while ensuring that all attending students had a seat (Table 2).

Table 2. Classroom organization and student distribution.

Class room	Max. Capacity	Group	Assigned Students	Estimated students by LR	Percentage of estimated attendance by LR	Estimated students by PR	Percentage of estimated attendance by PR
1	50	Reduced	52	43,73	87,46%	44,72	89,44%
2	50	Reduced	51	42,89	85,78%	43,86	87,72%
3	50	Reduced	53	44,57	89,15%	45,58	91,16%
4	120	Reduced	106	89,15	74,29%	91,16	75,97%
5	200	Complete	151	119,29	59,65%	120,95	60,48%
6	120	Complete	89	70,31	58,59%	71,29	59,41%
7	100	Complete	69	54,51	54,51%	55,27	55,27%
8	100	Complete	70	55,30	55,30%	56,07	56,07%
9	200	Complete	162	127,98	63,99%	129,76	64,88%
10	80	Absent	248	27,28	34,10%	28,20	35,25%
11	80	Absent	247	27,17	33,96%	28,08	35,10%
12	50	Absent	155	17,05	34,10%	17,62	35,25%
13	50	Absent	157	17,27	34,54%	17,85	35,70%
14	50	Absent	159	17,49	34,98%	18,08	36,16%

By distributing the students in this manner, we were able to effectively allocate the available resources and accommodate the predicted attendance levels for each group. The varying classroom capacities allowed for flexibility in accommodating different numbers of students within each group. The allocation of students to classrooms based on their respective groups ensured that the seating arrangements were optimised, and the available space was utilised efficiently. This approach aimed to strike a balance between maximizing capacity utilization and ensuring that all attending students had a seat. Overall, the results of this allocation strategy enabled the creation of an organised and conducive environment for the exam, where students had appropriate seating arrangements according to their attendance status. By achieving an optimal occupancy level (4) in each classroom, we were able to utilise the available space effectively while accommodating the expected number of attending students.

Table 3. Actual attendance of students and predictions from both estimators for the second exam.

Global Attendance	Percentage of global attendance	Linear Regression Global Attendance estimation	Linear Regression Attendance accuracy	Polynomial Degree 3 Global Attendance Estimation	Polynomial of degree 3 global attendance accuracy
721	40,57%	757,43	95,19%	774,90	93,04%

Table 4 shows the percentage of students that attended the exam, and the accuracy of each model. Note that the occupation level of all classrooms were between 38% and 84%, with a mean of 56,35% and a standard deviation of 13,28%. Even though both models predicted with high precision the attendance percentage with 2 3 percent difference (Table 3), classroom attendance distribution was not as accurate as expected. The lack of precision in the predictions can be attributed to the estimated assignment of probabilities to each student group. The approach used in this study involved assigning probabilities to achieve the overall predicted attendance, rather than predicting attendance rates for each group individually. However, data obtained from the second exam was used to calculate a much better probability of attendance for each group, as shown in Table 4, by adjusting the percentage of students who actually were present in each classroom combined with the group assigned to each classroom.

For the third exam, after data from the second exam was added to the dataset, a similar approach was taken. However, we decided to split the Reduced group into two subgroups, absent reduced (students that never attended any exams) and reduce (students that attended at least once). A different criteria was taken for the complete groups: we considered as complete and absent all students who never attended any exam or who attended only once and got insufficient. Another important consideration for the third exam is that not all classrooms were available: classrooms 7, 10 and 11 were occupied by other courses, so we had about 260 less seats available. However, since the LR model predicted an attendance rate of 33,35% of students and PR model predicted 35,75% attendance, this was not a real issue to attend. Results showed an actual attendance rate of 36,5%, so in this case PR model performed better than LR model (Table 4), with an error of only 0.75% (12.7 students over 600). For the allocation of students in classrooms we took a similar approach for the reduced exam students, but a different one from complete and absent: considering a near 1:2 relation for these groups (467 students for the complete group, and 976 for the absent group), we distributed one complete for every two absent in each classroom. To measure the estimation classroom occupation level, we calculated the expected value E for the whole complete group (complete and absent) as:

$$E = \frac{C \times P(C) + CA \times P(CA)}{C + A} \qquad (1)$$

where C is the total number of students in the Complete group, CA is the total number of students in the Complete Absent group, and P(C) and P(AC) are the probabilities for students in each group attending the exam. These probabilities are based on the estimation of both LR and PR models (Table 6).

Table 4. Accuracy of estimators in each classroom for the second exam. An accuracy of 100% means a perfect prediction, while an accuracy of 100% ± 25% means a fair enough prediction. Note that classrooms 10 to 14, in which students were absent in previous exams, tend to obtain less accurate predictions. However, even though the number of assigned students were very high, the occupation level was similar to most classrooms.

Classroom	Real attendance percentage	LR accuracy	PR accuracy
1	76,00%	86,89%	84,97%
2	68,00%	79,27%	77,52%
3	84,00%	94,23%	92,15%
4	66,67%	89,74%	87,76%
5	49,50%	82,99%	81,85%
6	52,50%	89,60%	88,37%
7	46,00%	84,39%	83,23%
8	42,00%	75,95%	74,91%
9	57,00%	89,08%	87,85%
10	61,25%	179,62%	173,77%
11	50,00%	147,22%	142,43%
12	38,00%	111,44%	107,81%
13	46,00%	133,18%	128,85%
14	52,00%	148,66%	143,82%

Table 5. Attendance rate per classification group after the second exam. The average classroom attendance rate column indicates a good distribution of students that ensures good attendance rate in all classrooms.

Group	Assigned of students	Actual students	Global attendance rate	Average classroom attendance rate
Reduced	262	194	74,04%	74,5%
Complete	541	364	67,28%	69,4%
Absent	966	158	16,35%	15%

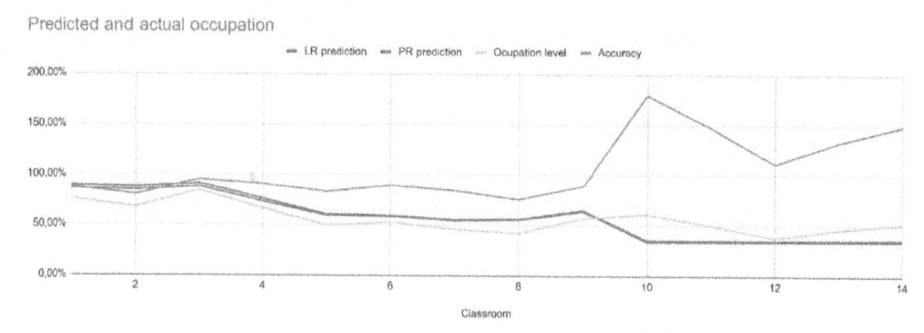

Fig. 1. Models' predictions and accuracy per classroom.

Table 6. Attendance estimations using the expected value E for both complete and complete absent groups.

Group	Number of students	LR Estimation	PR Estimation
Global estimation	1645	33,35%	35,75%
Reduced	282	83,00%	86,20%
Complete	467	78,10%	80,20%
Reduced Absent	60	4,90%	7,00%
Complete Absent	976	7,50%	9,00%
C + CA	1443	E = 28,59%	E = 30,69%

Table 7. Actual attendance of students and predictions from both estimators for the third exam

Global Attendance	Percentage of global attendance	Linear Regression Global Attendance estimation	Linear Regression Global Attendance accuracy	Polynomial degree 3 Global Attendance Estimation	Polynomial of degree 3 global attendance accuracy
600	36,50%	551,3	91,88%	587,31	97,88%

As mentioned above, even though the PR model performed better than the LR model for the third exam, both models performed accurately enough to make a good distribution of students, auxiliars and exams. Table 8 shows the accuracy levels achieved in each classroom, with a mean of 105,10% ± 62,67% std. Dev. And a median of 98,16%. The error rate value was higher than expected, nevertheless most classrooms were occupied between 40% and 80%, which fulfills the purpose of this work. We observed that the error rate was primarily due to classroom 13, which presents a special case in which both models predicted less than half of the students that actually attended. However, this situation was expected due to the uncertainty of the group classified as *reduced absent*, and it did not present a real issue since its occupancy level was 20% (10 students over 41, with predictions between 3,08 and 3,69) (Table 7).

Table 8. Actual attendance, models' estimation and models' estimation per. Classrooms 1, 2, 3 and 14 were assigned to reduced exams only, classroom 13 to the reduced absent group, and the rest for complete and complete absent groups with a 1:2 ratio.

Classroom	Attendance percentage	LR estimation	LR accuracy	PR estimation	PR accuracy
1	77,78%	73,78%	110,97%	76,62%	106,85%
2	84,44%	75,62%	111,67%	78,54%	107,52%
3	73,33%	71,93%	101,95%	74,71%	98,16%
4	75,00%	50,27%	151,34%	53,96%	140,99%
5	46,50%	53,03%	88,15%	56,93%	82,12%
6	78,33%	50,51%	155,82%	54,22%	145,16%
8	63,33%	57,75%	132,25%	61,99%	123,20%
9	44,50%	52,32%	83,24%	56,16%	77,54%
12	38,00%	42,88%	89,81%	46,03%	83,66%
13	20,00%	6,15%	325,20%	7,38%	271,00%
14	36,00%	66,40%	54,22%	68,96%	52,20%

A last prediction of the models was the rate of students that would pass the exam. In the second exam, the LR model predicted a 19,5% of approval rate (139 students), while the PR model predicted a 20,42% approval rate (146 students). The real approval rate was 18,88%, corresponding to 135 students, so even though the predictions weren't perfect, the models performed well enough to estimate the number of students who would pass the second exam. For the third exam, the LR model predicted a 15,93% approval rate, while the PR model predicted a 16,02% approval rate. In this case, the real approval rate was 19,01%, so the models performed less accurately in the third exam, but again close enough (with a difference of about 20 students). The precision of these early estimations are very useful to gain several weeks in advance to organise CADP for the second semester (retake format) and Programming Workshop, the course that follows CADP in the same semester.

5 Analysis and Conclusions

The primary objective of the attendance prediction models was to estimate the number of students expected to attend the exam, allowing for effective allocation of resources and seating arrangements. While the models may not have provided exact attendance figures, their predictions provided a reliable guideline for organizing the exam logistics. It is worth noting that although the predictions of the models were not exact, they were accurate enough to ensure an appropriate level of occupancy in all the classrooms used, avoiding situations where classrooms were overcrowded or nearly empty. LR model performed slightly better than PR model, both for attendance and approval estimations. However, as the dataset expands and more years of course data are incorporated, it is anticipated that the models will become increasingly accurate and robust in their predictions, and we expect the PR model to improve its accuracy by capturing the statistical variations between the both semesters.

By considering the predicted attendance levels, the allocation of students to classrooms could be carefully planned to achieve an optimal utilization of space: classrooms were neither overcrowded nor sparsely populated. This outcome is crucial in maintaining a smooth and efficient exam administration process while ensuring that all attending students had a seat available to them. While it is always desirable to have precise attendance predictions, the fact that the models provided sufficiently accurate estimates to avoid overcrowded or empty classrooms should be considered a strength of the approach.

One of the limitations of this study is that the assignment of attendance probabilities for different student groups was done in a way to achieve an overall attendance percentage according to the selected model. A future improvement could involve predicting not only the overall attendance but also the attendance percentages for each specific student group. This enhancement to the predictive models can be used to gain more granular insights into attendance patterns and better understand the dynamics within different student cohorts.

As noted before, the adjustments per group (Tables 5 and 8) were made by using *really* simple average approaches, with a small difference for the third exam in which we mixed students for the complete and absent groups. However, a much better adjustment could be achieved by running an optimization algorithm per group, such as minimum square error (MSE). For instance, let's consider attendance observations for the second exam, together with attendance predictions based on the LR model. We have already calculated the accuracy of the model in each classroom, and we also know the group each classroom belongs to (either reduced, complete or absent). Having this information into consideration, we can set a limitation per group using the real number of students which actually attended the exam, and calculate the best value of X for the same group (estimated attendance rate of the group X) by minimizing the square error between 1 (perfect estimation) and the model estimation per classroom.

Another improvement to this model would be implementing a classification model to categorise students into four possible groups: 1) Will pass, 2) Will fail, 3) Will receive an insufficient grade, and 4) Will not attend the exam. The classification model could be built using techniques such as logistic regression, decision trees, or support vector machines. By utilizing a classification model, we can go beyond predicting the overall pass rates and gain insights into the likelihood of individual students falling into different performance categories. This information can provide a more detailed understanding of student outcomes, and would enable educational institutions to identify students who may require additional support or intervention early on, facilitating timely interventions to improve their chances of success.

To implement this improvement, we would need to gather additional data that captures factors influencing student performance, such as prior academic records and engagement in coursework. One advantage of a classification model is the amount of available data: while regression models are based on less than 25 records (3 exams per semester, two semesters per year, 4 years of data plus data from this year), a classification model could base its predictions on several thousands records (between 500 and 3000 per semester). However, gathering records and constructing and preparing this dataset would require more work than the one used in this study.

References

1. Vissers, M.: Predicting students' class attendance. master thesis data science business and governance. Tilburg University, School of Humanities. Tilburg, The Netherlands, October 2018
2. Muzaferija, I., Mašetić, Z., Jukic, S., Kečo, D.: Student attendance pattern detection and prediction. J. Eng. Nat. Sci. **3** (2021). https://doi.org/10.14706/JONSAE2021313
3. Fernandes, E., Holanda, M., Victorino, M., Borges, V., Carvalho, R., Van Erven, G.: Educational data mining: predictive analysis of academic performance of public school students in the capital of Brazil. J. Bus. Res. **94**. https://doi.org/10.1016/j.jbusres.2018.02.012
4. Yağcı, M.: Educational data mining: prediction of students' academic performance using machine learning algorithms'. Smart Learn. Environ. **9**, 11 (2022). https://doi.org/10.1186/s40561-022-00192-z
5. Rashid, E., Ansari, M.D., Gunjan, V.K., Khan, M.: Enhancement in teaching quality methodology by predicting attendance using machine learning technique. In: Gunjan, V., Zurada, J., Raman, B., Gangadharan, G. (eds.) Modern Approaches in Machine Learning and Cognitive Science: A Walkthrough. Studies in Computational Intelligence, vol. 885, pp. 227–235. Springer, Cham (2020). https://doi.org/10.1007/978-3-030-38445-6_17
6. Reason, R.: Student variables that predict retention: recent research and new developments. J. Stud. Affairs Res. Pract. **40**(4), 704–723 (2003). https://doi.org/10.2202/1949-6605.1286
7. Credé, M., Roch, S.G., Kieszczynka, U.M.: Class attendance in college: a meta-analytic review of the relationship of class attendance with grades and student characteristics. Rev. Educ. Res. **80**(2), 272–295 (2010). https://doi.org/10.3102/0034654310362998
8. Villarreal, G.L.: Tasa de asistencia y aprobación a exámenes de CADP. Universidad Nacional de La Plata (2023). https://doi.org/10.35537/10915/157959

Computer Security

Computer Security

1.1 Common Network Security Issues

Computer Virus. Computer viruses are among the most common network threats in the field of cybersecurity. These are specifically designed to spread from one system to another. They can enter the user's computer in several different ways like: email attachments, downloads from the Internet, etc. Computer viruses are known for causing problems like: stealing of passwords or files, rendering computers inoperable, disabling security settings, etc.

Phishing. This type of attack usually impersonates institutions like banks in order to steal the user's sensitive data like passwords and credit card numbers. After opening the URL of the malicious website, the user is prompted to enter the financial details, which are then sent directly to the malicious source.

Denial of Service. In this type of attack, the hackers flood the networks with a large amount of traffic. This causes the system to slow down or completely paralyze. Once the network has been infiltrated, it can cause a number of issues for legitimate users like denying access to different types of services, preventing information from being retrieved from websites, etc.

Rogue Security Software. These attacks make the users believe that their system is facing a network security issue or that they need to update their system's security measures. Then they will make the user download their program to get the alleged malware removed. In this way, the actual virus is installed on the user's computer [1].

1.2 Ways to Prevent Network Attacks

Antivirus Software. This method is among the first lines of defense against different types of viruses. Antivirus softwares help in preventing malwares from being installed on the user's system. For efficient performance, these softwares should be updated regularly by the user.

Installing Firewall. Installing a firewall is one of the most efficient ways to protect the network against any kind of cyber-attack. The firewall obstructs any brute force attack made on the network or systems before it can do any damage. There are two types of firewalls: Host-based firewalls and Network-based firewalls.

Monitoring Network Activity. Tracking of network activities like logs and other data enables the timely detection of suspicious patterns or trends. This allows the concerned authorities to take the necessary steps to examine and provide protection against potential attacks.

Prediction of TCP Firewall Action Using Different Machine Learning Models

Amit Kumar Bairwa[1]($\boxtimes$), Akshit Kamboj[1], Sandeep Joshi[1], Pljonkin Anton Pavlovich[2], and Saroj Hiranwal[3]

[1] Manipal University Jaipur, Jaipur, India
amitkumar.bairwa@jaipur.manipal.edu
[2] Southern Federal University, Rostov-on-Don, Russia
[3] Victorian Institute of Technology, Adelaide, South Australia

Abstract. In today's world, the issues associated with network security have increased by a tremendous amount. For instance, cyber-attacks during different types of network transmissions are one of the major problems associated with security. A Firewall can be used to provide protection against unauthorized traffic during these transmissions. Our proposed solution uses several different machine learning techniques to predict the TCP firewall action on the basis of the transmission characterisitics of the TCP model. The main idea is to study different features of a TCP transmission like Source Port, Destination Port, Elapsed Time (in seconds), NAT Source Port, NAT Destination Port. Based on the analysis performed on these features, the TCP firewall action will be classified into one of the four categories. These categories are: Allow, Deny, Drop, Reset-Both. In this project, nine different machine learning models have been trained using an available dataset. The dataset used has over 65000 rows, where each row represents a TCP transmission. Each TCP transmission has been described by around 11 different features. A few examples of machine learning models that have been used include: Decision Tree Classifier, Random Forest Classifier, XGBoost Model, and Gradient Boosting Model.

Keywords: Steganography · SHA-256 · AES · LSB · Image

1 Introduction

Over the last decade, network security risks are on a continuous rise. Connecting systems and computers has provided several different benefits to businesses. However, the process of networking has also contributed significantly to the increase in the number of network security threats that different organizations face. Some of the most common problems include: loss of data, security breaches, and malicious attacks with the help of hacking and viruses. Therefore, it has become very important to take a number of precautions in order to reduce the vulnerability of these computer networks.

Data Encryption. In encryption, the important data is encoded in a secure manner, such that only authorized users have access to it. This helps in preventing network information from being compromised or stolen. There are several types of algorithms like AES, SHA 1, and MD5 that are used for encrypting and decrypting the data.

2 Related Works

The main aim of the research by Kimoon Han et al. [2] is to propose a Loss of Descrimination Algorithm (LDA) based on Machine Learning, for wireless TCP congestion control. A multi-layer perceptron model is used to train the LDA to distinguish packet losses due to the wireless channel environment and congestion. The congestion control will classify the cause of losses on the basis of the training results. The algorithm will not reduce the congestion window in the case of random losses. The algorithm has been implemented in the Linux kernel and a testbed has also been configured. This has been done to measure the performance of the proposed algorithm.

The research by Desta Haileselassie Hagos et al. [3] aims to show how an intermediate node like a network operator can be used for identifying the transmission state of the TCP client, with the help of the passive monitoring of the TCP traffic. A machine learning model is trained based on the inference obtained from the variant which underlies the loss-based TCP algorithms and Congestion Window. The same method can also be used for predicting the client's other TCP transmission states.

The primary aim of the research by Sang-Jin Seo et al. [4] is to enhance the improvement of the TCP congestion control algorithm v2, which is based on the Deep Q Network (DQN), in a single flow. Another major goal of this work is to improve the inter-protocol fairness when the same bottleneck link is being shared among several flows. All of this has been implemented using Reinforecment Learning Technique.

The main aim of the research by I.L. Afonin et el. [5] is to determine packet delay time and network load using a machine learning algorithm with reinforcement. This algorithm makes use of TCP CUBIC. The algorithm observes the behaviour of the window overload size and learns from it. Based on these learnings, the algorithm makes optimal decisions using the reward function. Improvements in the performance are observed from the simulation results. These improvements include: reduced packet latency and increased bandwidth as compared to various exisiting TCP algorithms.

The research by Kimoon Han et al. [6] aims to distinguish wireless errors and network congestions using a deep learning algorithm. This distinction should take place when a packet loss occurs. Congestion control is executed in the same manner as existing TCP in the case of loss caused by network congestion. On the other hand, an algorithm has been proposed in the case of loss due to errors. This algorithm aims to improve the wireless TCP performance by just retransmitting the packets that have been lost without reducing the congestion window.

The primary aim of the research by Desta Haileselassie Hagos et al. [7] is to propose and evaluate an advanced classification approach to the passive operating system (OS) fingerprinting. This is done using several different machine learning and deep learning techniques. For this research, controlled experiments have been conducted on bencmark data, realistic, and emulated traffic. In the approach using Oracle-based machine learning, the underlying TCP variant is an important feature for the prediction of the remote OS. On the basis of this observation, a very sophisticated tool has been developed for OS fingerprinting. This tool first predicts the TCP flavor by making use of passive traffic traces. This prediction in turn is used as an input feature by the tool for the prediction of the remote OS from the passive measurements.

The research by Ning Li et al. [8] aims to propose a congestion control strategy named 'AdaBoost-TCP.' This is a congestion control strategy that is based on machine learning. In AdaBoost-TCP, an adaptive Boost recognition model has been constructed. This model can efficiently classify the type of packet loss in the given satellite network. This enables the sender to adopt the appropriate congestion control measures based on the packet loss type. If the network load is not increased, then the AdaBoost-TCP can have higher efficiency and better classification speed.

The main aim of the research by P. Geurts et al. [9] is to propose the application of several machine learning techniques in improving TCP congestion control in both wireless and wired networks. Thus, the main goal is to build a loss classifier based on the dataset obtained by performing simulations of random network topologies. For this task, several machine learning algorithms have been used and compared. Decision tree boosting is the best method for this application.

The primary aim of the research by Quanling Zhao et al. [10] is to make predictions related to TCP Firewall action using different machine learning techniques. The prediction is based on the characteristics of TCP transmission. For this purpose, several different machine learning models have been used like support vector machine (SVM), neural network, logistic regression, and AdaBoost. A final accuracy of over 98% has been achieved using the concept of ensemble-learning.

3 Statement of the Problem

"Prediction of TCP Firewall action using different machine learning algorithms. Comparing these models based on parameters such as accuracy, efficiency. Finally, after proper comparison, determining the best model."

4 Objectives of the Study

In today's world, the threat posed by cyber-attacks has increased by a tremendous amount. These attacks can cause several problems like data breaches, injection of malware into the users' devices, and denial of services. This makes it very essential to provide appropriate protection against these types of attacks.

Through this study, efficient machine learning models can be built for predicting the TCP Firewall action for different types of TCP transmissions. The main objective of this research work is to gain valuable insights from several different research papers available at various conferences and journals. These papers will be closely related to the kind of work defined in our problem statement. This will enable us to learn about various kinds of techniques that can be used to tackle the given problem.

5 Requirements

5.1 System Architecture

1. RAM: Minimum 8–16 Gb
2. CPU: Intel Corei5 7th Generation or more is preferred.
3. Storage: Minimum 128 GB or more.
4. OS: Linux/Windows/macOS
5. Internet: High speed internet connectivity required.

5.2 Technology Used

1. Python
2. Sckit-Learn
3. Pandas
4. Matplotlib
5. Jupyter Notebook Or Google Colab
6. Basic Knowledge Of Machine Learning And Data Science

6 Data Preprocessing

6.1 Preparing the Dataset

For this work, a TCP transmissions dataset has been used. This dataset has been obtained from the UCI machine learning repository. The dataset provides information about characteristics of different types of TCP transmissions. There are over 65000 rows (data points) and 11 columns (features) in the given dataset. Information related to different features in the dataset is displayed in the Table 1.

6.2 Exploratory Data Analysis

This step involved getting utterly familiar with the dataset. Various aspects of the dataset have been studied [11]. In this stage, the following tasks have been performed:

1. Exploration of selected features to identify trends and patterns.
2. Re-ordering the columns present in the dataset.
3. Studying the distribution of different target classes (as shown in Fig. 1).

Table 1. Different features in the Dataset

Index	Column	Non-Null Count	Dtype
0	Source Port	65532 non-null	int64
1	Destination Port	65532 non-null	int64
2	NAT Source Port	65532 non-null	int64
3	NAT Destination Port	65532 non-null	int64
4	Action	65532 non-null	object
5	Bytes	65532 non-null	int64
6	Bytes Sent	65532 non-null	int64
7	Bytes Received	65532 non-null	int64
8	Packets	65532 non-null	int64
9	Elapsed Time (sec.)	65532 non-null	int64
10	$pkts_sent$	65532 non-null	int64
11	$pkts_received$	65532 non-null	int64

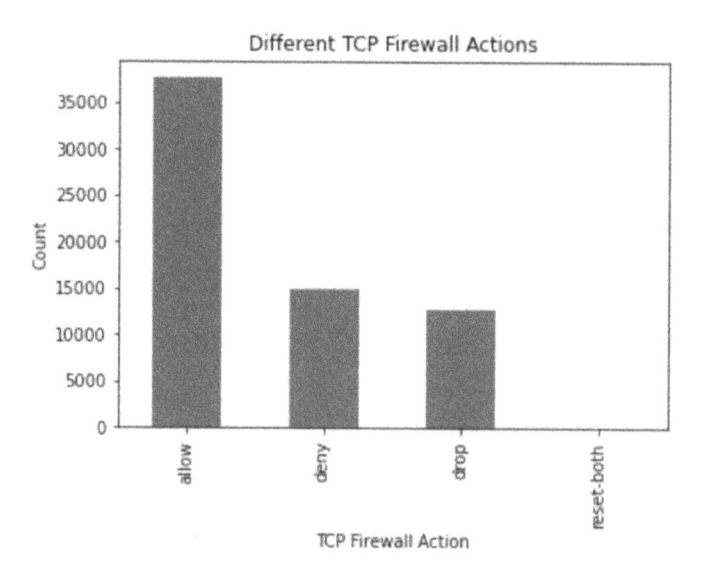

Fig. 1. Different TCP Firewall Actions

4. Converting the categorical data into numerical data.
5. Balancing the dataset using Upsampling technique.
6. Scaling the different features in the dataset.
7. Creating copies of the prepared dataset at regular intervals.
8. Dividing the dataset based on dependent and independent variables.
9. Performing the Train-Test split on the dataset.

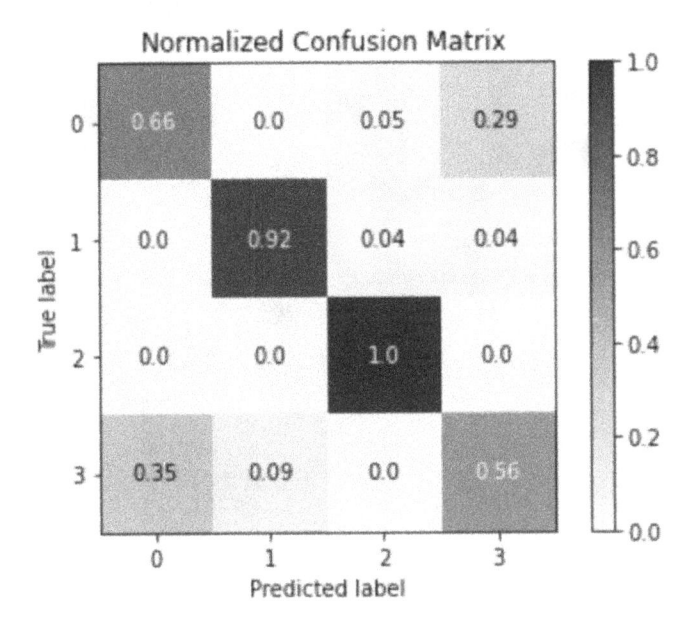

Fig. 2. Confusion Matrix: Logistic Regression

7 Implementation of Machine Learning Models

7.1 Logistic Regression

An accuracy of 78.77% has been achieved so far on the Training Dataset. On the other hand, an accuracy of 78.57% has been achieved on the Test Dataset. The confusion matrix for the Logistic Regression model after normalization is shown in the Fig. 2.

7.2 Decision Tree Classifier

An accuracy of 99.96% has been achieved so far on the Training Dataset. On the other hand, an accuracy of 99.93% has been achieved on the Test Dataset. The confusion matrix for the Decision Tree model after normalization is shown in the Fig. 3.

7.3 Random Forest Classifier

An accuracy of 99.0% has been achieved so far on the Training Dataset. On the other hand, an accuracy of 98.98% has been achieved on the Test Dataset. The confusion matrix for the Random Forest model after normalization is shown in the Fig. 4.

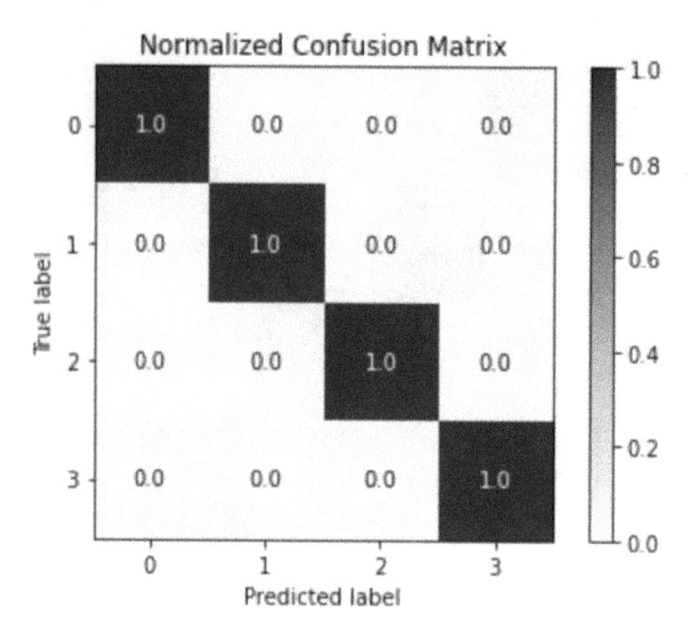

Fig. 3. Confusion Matrix: Decision Tree Classifier

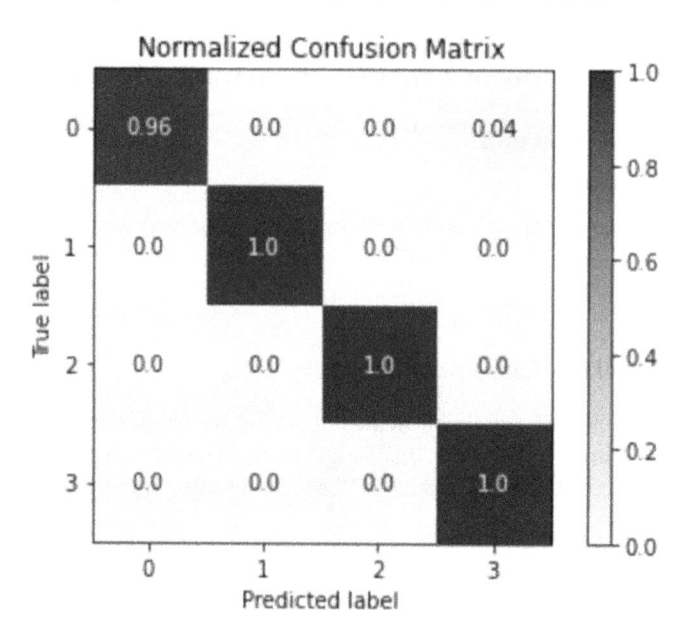

Fig. 4. Confusion Matrix: Random Forest Classifier

Prediction of TCP Firewall Action Using Different Machine Learning Models   169

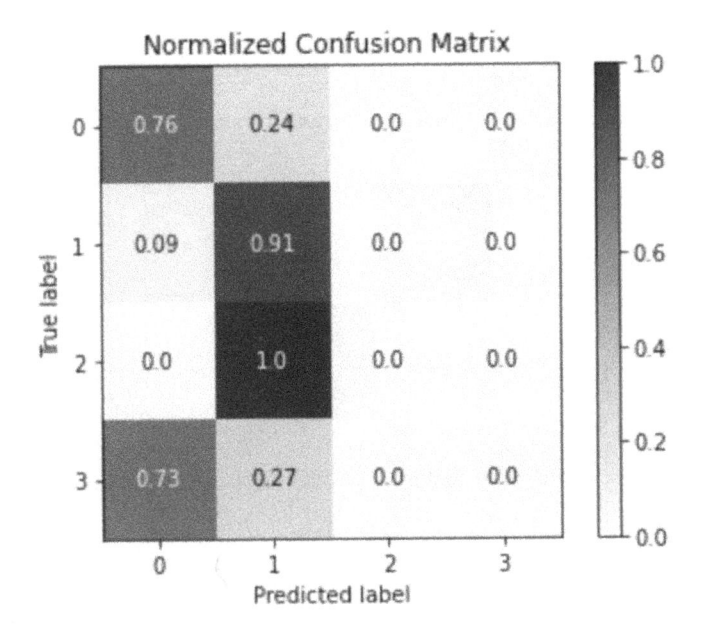

Fig. 5. Confusion Matrix: K-Means Clustering

7.4 K-Means Clustering

An accuracy of 42.06% has been achieved so far on the Training Dataset. On the other hand, an accuracy of 42.04% has been achieved on the Test Dataset. The confusion matrix for the K-Means model after normalization is shown in the Fig. 5.

7.5 Principle Component Analysis

An accuracy of 78.85% has been achieved so far on the Training Dataset. On the other hand, an accuracy of 78.62% has been achieved on the Test Dataset. The confusion matrix for the Principle Component Analysis model after normalization is shown in the Fig. 6.

7.6 Support Vector Machine

An accuracy of 74.41% has been achieved so far on the Training Dataset. On the other hand, an accuracy of 74.05% has been achieved on the Test Dataset. The confusion matrix for the Support Vector Machine model after normalization is shown in the Fig. 7.

7.7 XGBoost Model

An accuracy of 99.97% has been achieved so far on the Training Dataset. On the other hand, an accuracy of 99.94% has been achieved on the Test Dataset.

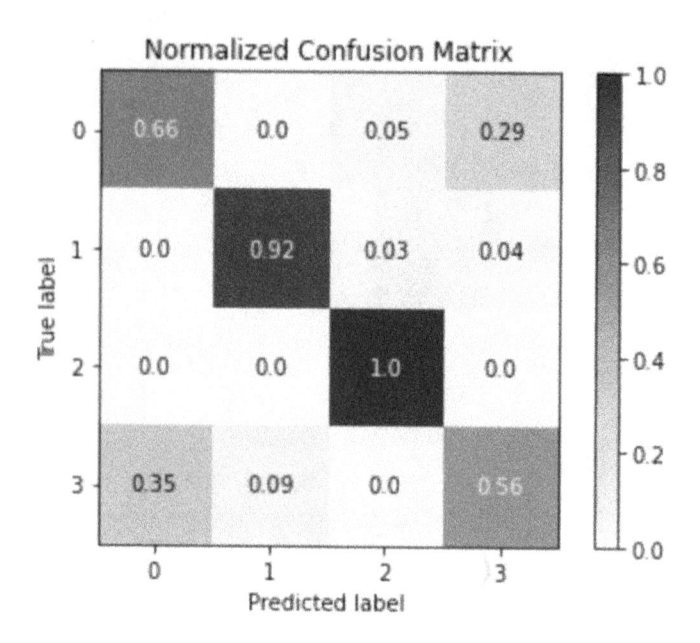

Fig. 6. Confusion Matrix: Principle Component Analysis

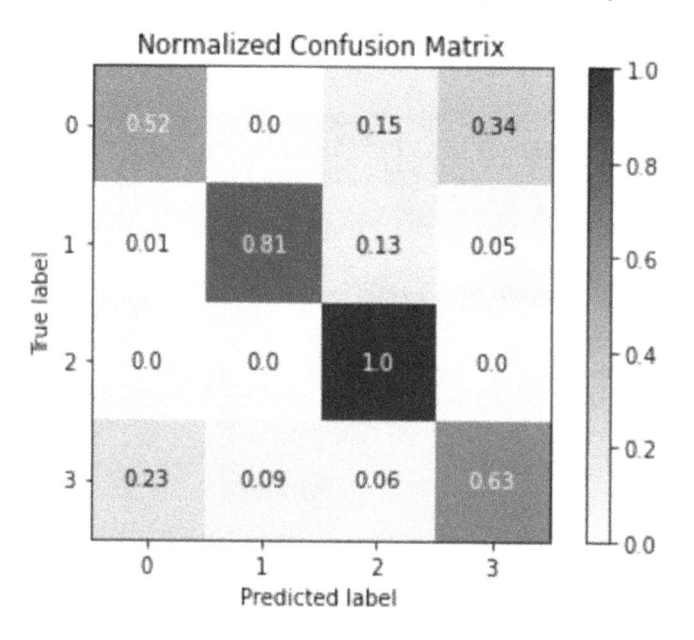

Fig. 7. Confusion Matrix: Support Vector Machine

The confusion matrix for the XGBoost model after normalization is shown in the Fig. 8.

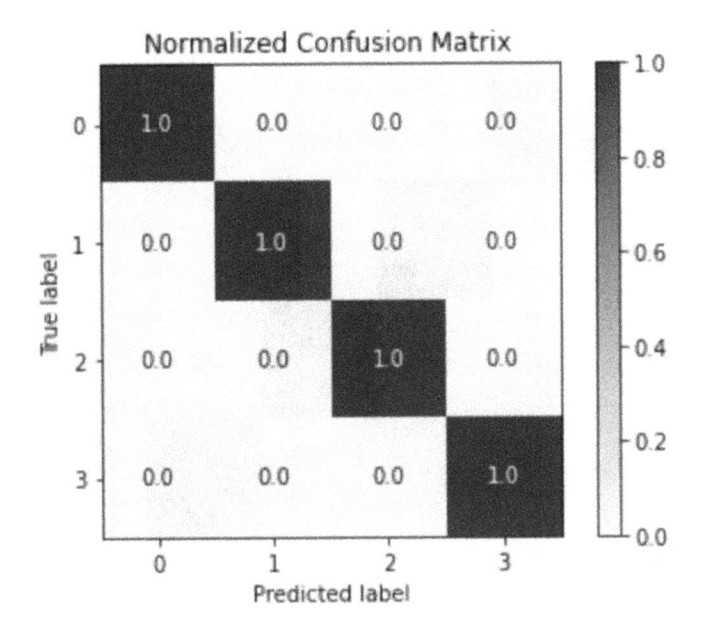

Fig. 8. Confusion Matrix: XGBoost Model

7.8 Gradient Boosting Model

An accuracy of 99.64% has been achieved so far on the Training Dataset. On the other hand, an accuracy of 99.61% has been achieved on the Test Dataset. The confusion matrix for the Gradient Boosting model after normalization is shown in the Fig. 9.

7.9 AdaBoost Classifier

An accuracy of 68.7% has been achieved so far on the Training Dataset. On the other hand, an accuracy of 69.07% has been achieved on the Test Dataset. The confusion matrix for the AdaBoost model after normalization is shown in the Fig. 10.

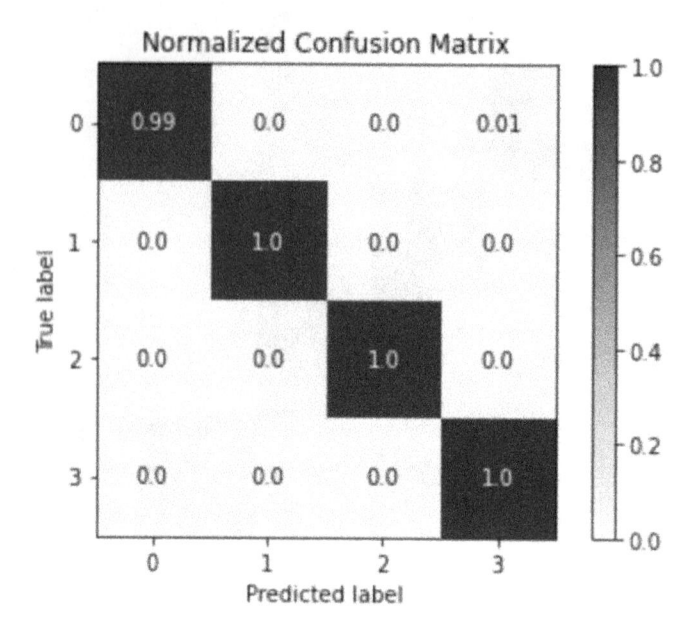

Fig. 9. Confusion Matrix: Gradient Boosting Model

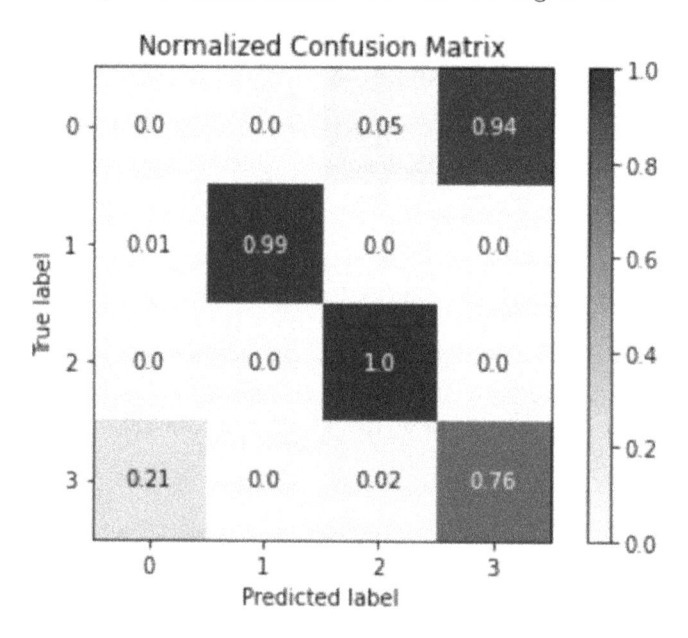

Fig. 10. Confusion Matrix: AdaBoost Classifier

8 Conclusion

A large number of tasks have been performed while working on this research project. The first task was to formulate a problem statement which is required to be solved. After this, a large number of research papers (closely related to the problem statement) were put into analysis. The step after this involved finding an adequate dataset. The dataset has been obtained from the UCI machine learning repository. This obtained dataset consists of various important features which have a great influence on deciding the appropriate TCP Firewall action. After preparing the dataset, it was further explored and several changes were incorporated into it. Then Train-Test Splits were performed on the modified dataset. Finally, the training dataset was fetched to nine different Machine Learning Models like: Logistic Regression, Decision Tree, and Random Forest Model. Out of the these nine models, the K-Means Clustering Model has been observed to have the least accuracy. On the other hand, extremely high accuracies were observed in the case of Decision Tree, Random Forest, XGBoost, and Gradient Boosting Models.

9 Future Work

In the case of our project, a large number of ideas can be implemented both in the near future and over a long term. In the immediate future, the primary goal will be to keep trying different Machine Learning Models and determine the best model as per the needs. At the same time, there is also a need to make improvements in the prediction accuracy of each model tried. There is also a need to take care of several things like avoiding multicollinearity, avoiding model overfitting, and so on. This is especially true in the case of Decision Tree and Random Forest Models. During these different processes, the main priority will always be to achieve the highest accuracy while making sure that problems like multicollinearity and overfitting do not occur.

References

1. Detection of malware in downloaded files using various machine learning models. Egypt. Inform. J. **24**(1), 81–94 (2023)
2. Han, K., Lee, J.Y., Kim, B.C.: Machine-learning based loss discrimination algorithm for wireless TCP congestion control. In: 2019 International Conference on Electronics, Information, and Communication (ICEIC), pp. 1–2 (2019)
3. Hagos, D.H., Engelstad, P.E., Yazidi, A., Kure, Ø.: General TCP state inference model from passive measurements using machine learning techniques. IEEE Access **6**, 28372–28387 (2018)
4. Seo, S.-J., Cho, Y.Z.: Fairness enhancement of TCP congestion control using reinforcement learning. In: 2022 International Conference on Artificial Intelligence in Information and Communication (ICAIIC), pp. 288–291 (2022)

5. Afonin, I.L., Gorelik, A.V., Muratchaev, S.S., Volkov, A.S., Morozov, E.K.: Development of an adaptive TCP algorithm based on machine learning in telecommunication networks. In: 2019 Systems of Signal Synchronization, Generating and Processing in Telecommunications (SYNCHROINFO), pp. 1–5 (2019)
6. Han, K., Hwang, A., Lee, J.Y., Kim, B.C.: Design and performance evaluation of enhanced congestion control algorithm for wireless TCP by using a deep learning. In: 2018 International Conference on Electronics, Information, and Communication (ICEIC), pp. 1–2 (2018)
7. Hagos, D.H., Yazidi, A., Kure, Ø., Engelstad, P.E.: A machine-learning-based tool for passive os fingerprinting with tcp variant as a novel feature. IEEE Internet Things J. **8**(5), 3534–3553 (2021)
8. Li, N., Deng, Z., Zhu, Q., Du, Q.: AdaBoost-TCP: a machine learning-based congestion control method for satellite networks. In: 2019 IEEE 19th International Conference on Communication Technology (ICCT), pp. 1126–1129 (2019)
9. Geurts, P., El Khayat, I., Leduc, G.: A machine learning approach to improve congestion control over wireless computer networks. In: Fourth IEEE International Conference on Data Mining (ICDM'04), pp. 383–386 (2004)
10. Zhao, Q., Sun, J., Ren, H., Sun, G.: Machine-learning based TCP security action prediction. In: 2020 5th International Conference on Mechanical, Control and Computer Engineering (ICMCCE), pp. 1329–1333 (2020)
11. Ahmadi, M., Ulyanov, D., Semenov, S., Trofimov, M., Giacinto, G.: Novel feature extraction, selection and fusion for effective malware family classification. In: Proceedings of the Sixth ACM Conference on Data and Application Security and Privacy, CODASPY '16, pp. 183–194, New York, NY, USA, 2016. Association for Computing Machinery (2016)

Author Index

A
Ahon, Santiago Ponte 45
Aidelman, Yael 59
Ansari Dogaheh, Morteza 103
Artola, Verónica 145

B
Bairwa, Amit Kumar 161
Bianco, Pedro Dal 45
Borrelli, Franco M. 30

C
Challiol, Cecilia 30

D
Dal Bianco, Pedro 59
De Giusti, Armando 3, 93
de Guzmán, Ignacio García-Rodríguez 121
de la Barrera Amo, Antonio García 121

E
Epelde, Francisco 103

G
Gamen, Roberto 59

H
Hasperué, Waldo 45, 59
Hiranwal, Saroj 161

J
Joshi, Sandeep 161

K
Kamboj, Akshit 161

L
Leon, Betzabeth 74
Luque, Emilio 74, 103

M
Mendez, Sandra 74
Moroni, Alejandro David 17

N
Naiouf, Marcelo 3

P
Palumbo, Maria Laura 17
Párraga Aranda, Jesús 132
Parraga, Edixon 74
Pavlovich, Pljonkin Anton 161
Pérez del Castillo, Ricardo 132
Pérez, Gabriel 17
Piattini, Mario 121, 132
Polo, Macario 121
Ponte Ahón, Santiago Andres 59
Pousa, Adrián 3

Q
Quiroga, Facundo 45, 59

R
Rexachs, Dolores 74, 103
Rios, Gastón 45
Ronchetti, Franco 45, 59
Russo, Claudia Cecilia 17

S
Sanz, Victoria 3
Seery, Juan Martín 59

Serrano, Manuel Ángel 121
Stanchi, Oscar 45, 59
Suppi, Remo 74

T
Taboada, Manel 103

V
Villarreal, Gonzalo L. 145

W
Walas Mateo, Federico 93
Wong, Alvaro 103

Printed in the USA
CPSIA information can be obtained
at www.ICGtesting.com
CBHW051749271024
16493CB00004B/85